Assessment Methods in Statistical Education

Assessment Methods in Statistical Education

An International Perspective

Edited by

Penelope Bidgood

Kingston University, UK

Neville Hunt

Coventry University, UK

Flavia Jolliffe

University of Kent, UK

WILEY

A John Wiley and Sons, Ltd., Publication

This edition first published 2010

Registered office
John Wiley & Sons Ltd, The Atrium, Southern Gate, Chichester, West Sussex, PO19 8SQ, United Kingdom

For details of our global editorial offices, for customer services and for information about how to apply for permission to reuse the copyright material in this book please see our website at www.wiley.com.

Library of Congress Cataloguing-in-Publication Data

Assessment methods in statistical education: an international perspective / edited by Penelope Bidgood, Neville Hunt, Flavia Jolliffe.
p. cm.
Includes bibliographical references and index.
ISBN 978-0-470-74532-8 (pbk.)
1. Mathematical statistics – Study and teaching – Evaluation. I. Bidgood, Penelope. II. Hunt, Neville. III. Jolliffe, F. R. (Flavia R.), 1942-
QA276.18.A785 2010
519.5071 – dc22

2010000192

A catalogue record for this book is available from the British Library.

ISBN 978-0-470-74532-8 (P/B)

Typeset in 10/12 Times-Roman by Laserwords Private Limited, Chennai, India
Printed and bound in Great Britain by TJ International Ltd, Padstow, Cornwall.

Contents

Contributors

Penelope Bidgood Faculty of CISM, Kingston University, UK. bidgood@kingston.ac.uk

Stephanie Budgett Department of Statistics, The University of Auckland, New Zealand. s.budgett@auckland.ac.nz

Neville Davies Faculty of Education, University of Plymouth, UK. neville.davies@rsscse.org.uk

María del Carmen Fabrizio Facultad de Agronomia, Universidad de Buenos Aires, Argentina. fabrizio@agro.uba.ar

Robert delMas Department of Educational Psychology, University of Minnesota, USA. delma001@umn.edu

Mike Forster Department of Statistics, The University of Auckland, New Zealand. m.forster@auckland.ac.nz

Joan Garfield Department of Educational Psychology, University of Minnesota, USA. jbg@umn.edu

Paula Griffiths Department of Human Sciences, Loughborough University, UK. p.griffiths@lboro.ac.uk

Ailish Hannigan Department of Mathematics and Statistics, University of Limerick, Ireland. ailish.hannigan@ul.ie

Alistair Harvey Department of Psychology, University of Winchester, UK. alistair.harvey@winchester.ac.uk

Neville Hunt Department of Mathematics, Statistics and Engineering Science, Coventry University, UK. n.hunt@coventry.ac.uk

Flavia Jolliffe Institute of Mathematics, Statistics and Actuarial Science, University of Kent, UK. f.jolliffe@kent.ac.uk

María Virginia López Facultad de Agronomia, Universidad de Buenos Aires, Argentina. mvlopez@agro.uba.ar

Helen MacGillivray Mathematical Sciences, Faculty of Science and Technology, Queensland University of Technology, Australia. h.macgillivray@qut.edu.au

Moira Maguire Department of Nursing, Midwifery and Health Studies, Dundalk Institute of Technology, Ireland. moira.maguire@dkit.ie

John Marriott School of Computing and Mathematics, Nottingham Trent University, UK. john@jmarriott.co.uk

Houshang Mashhoudy Department of Mathematics, Statistics and Engineering Science, Coventry University, UK. h.mashhoudy@coventry.ac.uk

Pam McKinney CILASS, Information Commons, UK. p.mckinney@sheffield.ac.uk

Rosie McNiece Faculty of CISM, Kingston University, UK. r.mcniece@kingston.ac.uk

Maxine Pfannkuch Department of Statistics, The University of Auckland, New Zealand. m.pfannkuch@stat.auckland.ac.nz

María Cristina Plencovich Facultad de Agronomia, Universidad de Buenos Aires, Argentina. plencovi@agro.uba.ar

Richard Rowe Department of Psychology, University of Sheffield, UK. r.rowe@sheffield.ac.uk

Milo Schield Department of Business Administration, Augsburg College, USA. milo@pro-ns.net

Zoë Sheppard Department of Human Sciences, Loughborough University, UK. z.a.sheppard@lboro.ac.uk

Vanessa Simonite School of Technology, Oxford Brookes University, UK. vsimonite@brookes.ac.uk

Rosemary Snelgar Department of Psychology, University of Westminster, UK. r.snelgar@westminster.ac.uk

Neil Spencer Business School, University of Hertfordshire, UK. n.h.spencer@herts.ac.uk

Doug Stirling Institute of Fundamental Sciences, Massey University, New Zealand. d.stirling@massey.ac.nz

Ralph Targett School of Technology, Oxford Brookes University, UK. rtargett@brookes.ac.uk

Sidney Tyrrell Department of Mathematics, Statistics and Engineering Science, Coventry University, UK. s.tyrrell@coventry.ac.uk

Mike Van Duuren Department of Psychology, University of Winchester, UK. mike.vanduuren@winchester.ac.uk

Chris J. Wild Department of Statistics, The University of Auckland, New Zealand. c.wild@auckland.ac.nz

Jamie Wood CILASS, Information Commons, UK. jamie.wood@sheffield.ac.uk

Andrew Zieffler Department of Educational Psychology, University of Minnesota, USA. zief0002@umn.edu

Foreword

In education, assessment is amongst the most useful things that we do for ourselves and our students. It is also amongst the most harmful things we do – the best and the worst.

It is useful for our students when it enables them to see what they do not understand and gives them insight and motivation to improve. It is useful for us as teachers when it helps us see where our teaching can be improved, when it gives us insight into the way our students are learning and when we can see the rewards of a job well done. It is useful for administrators when it helps them see which sort of structures work best for learning and which sort of people make good teachers, and ways in which they can improve the overall learning process.

It is harmful when it is seen as an end in itself. It is harmful to students when it makes the goal getting a paper qualification rather than gaining competence. It is harmful when it distorts the learning process and encourages learning and teaching for the test. Assessment is harmful when its contents do not match up with what is important to learn. To quote a phrase I first heard from Professor Hugh Burkhardt of the Shell Centre for Mathematical Education in Nottingham, 'what you test is what you get' – WYTIWYG. It is harmful when it is seen merely as a hurdle and when it promotes fear of failure, so encouraging strategies of getting high scores (particularly 'passing' an examination) at the expense of improving teaching and learning.

The position is made more difficult by the fact that many students studying statistics are not doing so out of choice. They may have to take a basic statistics course because it is an integral part of their main discipline – and they are not necessarily convinced of its usefulness. They may see it as an imposition, not an interesting learning experience to be applied in their profession. This makes it all the more likely that they will do the minimum necessary to get a piece of paper saying they have qualified.

All of the above may appear to say: formative assessment good, summative assessment bad. But it is not as easy as this. It is possible to develop good methods of summative assessment. This is only done by maintaining the focus that all assessment is subservient to the overall aims of improving teaching and learning and improving the statistical abilities of all our students.

In their different ways, the authors of this book explore the dilemmas posed by the need for good, relevant assessment and the wide variety of backgrounds and motives of students of statistics. It is interesting to compare the chapters of this book with those of the earlier book edited by Gal and Garfield (1997) to see how much work has been put in this area over the intervening 10 years. Things will continue to improve as we learn from each others' experiences and develop new ideas and methods.

Peter Holmes
Sheffield, UK

Preface

In 2007, the editors obtained funding from the MSOR Network of the UK Higher Education Academy and the Royal Statistical Society Centre for Statistical Education, to engage in the Variety in Statistics Assessment (ViSA) project. This project aimed to gather and disseminate evidence of successful experiences of assessment from teachers of statistics at tertiary level both within the UK and around the world. Although a number of meetings were convened and conference presentations given, the main focus of the project was the compilation of this book. Towards the end of 2007, a call for contributions was made through various publications and electronic news groups. The final accepted papers form the chapters of this book, which have been arranged into four themed parts.

One of the interesting features of statistics is that it is taught both as a specialist subject, often closely related to mathematics, and also within many other disciplines. Part of the value of this book is the fact that it draws on the experience not only of statisticians but also of educators from within subjects such as psychology, biology, business, health and agriculture. That richness is further enhanced by its international perspective, with authors drawn from six different countries across the world.

Students vary in their abilities and in their approach to learning. It is therefore only fair that the assessment process should allow a variety of opportunities for students to demonstrate their achievement. One of the key roles of the statistician is communication, explaining the results of often complex analysis to a client with little knowledge of the subject. The modern statistics lecturer must take this into account when devising an assessment strategy for a course of study. This theme is explored in Part A, where some authors outline general principles and others recount personal experience of successful assessment strategies.

Within the statistical education community, there is an ongoing debate about what is the essence of the subject. In an age where computers can not only perform all the necessary calculations, but also suggest an appropriate method of analysis and even write an automated report, many would argue that the development of statistical thinking is paramount. Assessing statistical thinking is much more difficult than assessing the ability to perform routine calculations. This is the challenge addressed by the authors in Part B.

Whilst theoretical statistics continues to be a thriving area of research, the vast majority of statistics teaching is in the applied area. Unsurprisingly, the focus of most statistics assessment is therefore on how what is learned relates to and enriches the world around us. Some examples of this 'real-world' assessment are found in Part C.

The role of technology in assessment has become increasingly important. On the positive side, the Internet has provided access to a wide variety of data sources, while the extensive availability of modern statistical computing packages allows lecturers to set more realistic tasks for student assignments. The negative aspect of technology is that it has facilitated plagiarism and collusion. Part D contains several accounts of how lecturers have responded to this threat to the integrity of the assessment process by developing techniques for individualised assessment.

Acknowledgements

The editors are grateful for the assistance of the following referees, whose knowledge and insight was invaluable in preparing the contributions to this book: Beth Chance, Neville Davies, David Goda, Gerald Goodall, Peter Holmes, John Marriott, Ann Ooms, Gilberte Schuyten and Larry Weldon.

Part A

SUCCESSFUL ASSESSMENT STRATEGIES

1

Assessment and feedback in statistics

Neville Davies and John Marriott

1.1 Introduction

Statistics is not only a subject in its own right but is also applied to diverse other subjects, including the sciences, geography, psychology, business and economics. Consequently assessment may need to cover a very wide-ranging set of topics and activities taught within a statistics curriculum. In this chapter we consider the implications of the ubiquity of the application of statistics and its pedagogy, the significant impact of this broad 'learning base' on what should be assessed and the plethora of ways assessment and feedback can be provided.

1.2 Types and purposes of assessment

There are four types of assessment that we identify in relation to statistics:

1. *Diagnostic assessment* seeks to identify the starting position of students, to identify gaps and thus enables these to be filled.
2. *Formative Assessment* seeks to use assessment for improvement, to indicate strengths and weaknesses and to give both student and teacher insight into the progress being made; formative assessment can also contribute some marks to overall assessment.
3. *Summative assessment* seeks to evaluate overall achievement, usually at the end of the course. With mid-course summative assessment it is possible to use the resulting feedback for formative purposes.

Assessment Methods in Statistical Education Edited by Penelope Bidgood, Neville Hunt and Flavia Jolliffe

4. *Evaluative assessment* is the final putting together of results and it involves the whole assessment of the programme of study. Its aims include the satisfaction of society, the institution, the teachers and, hopefully, the candidate. Such assessment covers the quality assessment of the course itself.

It is common sense that good assessment will lead to both good learning and good teaching, thereby encouraging a balanced view of the subject. MacGillivray argues in Chapter 2, the case that 'assessment is for learning' and that this is promoted by variety in statistics assessment. In the past there has been an emphasis on summative assessment. There is, however, an increasing recognition that only having assessment at the end of a course in statistics helps neither students nor staff. An increasingly varied approach to assessment has evolved in recent years, together with emphasis on feedback to students and staff. Bingham (2001) shows how assessment fits in with the whole teaching process. He clearly shows the interrelationships between learning outcomes, strategies for learning and teaching, assessment strategies and criteria for assessment, arguing cogently that assessment is an integral part of the whole process of teaching and learning.

Statistics, a complex subject with many components, provides the paradigm and methodology for making sense of data arising from a wide range of subjects. This means that we need to think long and hard about how we assess and provide feedback to students of statistics. The key reasons for doing assessment include enabling both students and their teachers to realise what they know, the teachers to know their own effectiveness, the institution to award grades to students, and society to judge the effectiveness of the awarding institution. No one, simple technique is going to provide all of this information. Thus, we take a broad view of the many methods and approaches available for assessment and feedback, but emphasise those most relevant for the statistics teacher.

1.3 What are we assessing?

Ideally an examiner will have been involved in designing the learning outcomes and assessment criteria of the course documentation. The assessment would then be constructed to determine reliably whether the learning outcomes have been met. In a statistical context there will inevitably be a range of topics requiring assessment. These follow from the fact that statistics as a subject, provides a paradigm for problem solving and is a methodology for doing science and many other subjects. Thus, the assessment of statistics should include assessment of:

- Understanding the statistical problem-solving approach (PSA).
- The course content, for example technical aspects of statistics.
- The process of doing statistics, for example undertaking an analysis that may include estimating parameters.
- The use of global skills, for example the use of appropriate statistical software.

- Personal and transferable (key) skills, for example skills inherent in statistical problem solving.
- Critical abilities such as those required in looking at the practical aspects of surveys.
- Communication of statistical results and conclusions.

In addition, the learning outcomes and their assessment need to take note of the levels at which learning in general and statistical learning in particular operate, and ensure a reasonable coverage.

Before looking at the positive aspects of assessment it is worth noting some of the problems that can occur in carrying it out. Brown (2001) provides a useful list of these problems, which includes: mismatches between the stated outcomes and the work required of the students; unclear or unstated assessment criteria; overuse of a particular form of assessment; and issues concerning feedback to students.

1.4 Formative assessment

Formative assessment of the student's work in statistics enables appropriate action to be taken by both student and teacher to correct misunderstandings. Black and Wiliam (1998) showed that effective feedback was a major means of developing learning. For this reason formative assessment is essential for students' involvement, motivation and understanding of their own progress and ways forward.

Successful formative assessment will employ 'resource-light' methods yet provide reliable information. Wider use of online testing and feedback has opened new ways of dealing with such assessment and some universities now make significant use of computer-based testing and feedback. Examples of these include online diagnostic testing and broader based tests that progress through a course with different levels of difficulty. Formative evaluation is often informal, but it can be formalised, for example by using short student-marked tests in class time or past papers for revision.

There is the problem that sometimes only the better students undertake work that does not count for their examination 'final' marks. Some statistics courses cope with this by awarding a very small proportion of the summative marks for all work done, including the formative assessment. Another approach is to get students to keep a log of all their formative work, some of which is later randomly selected for summative assessment. Sometimes marks are given for just doing the quiz or assessment, irrespective of the results. Logbooks are particularly useful for assessing the practice of statistics.

Marks, in general, may not be appropriate in formative assessment. Although good marks may give some encouragement, anecdotal evidence suggests that marks can detract students from looking carefully at their work. Marks give a sense of closure, whereas the purpose of formative assessment is to enable students to fill in the gaps and deepen their understanding. Hence discursive feedback comprising, for example, comments on suggested reading and reference to other class work, is essential.

There needs to be a relationship of trust between student and teacher and the formative evaluation must give a constructive and honest judgement of the work done. To be most effective, formative evaluation must be ongoing and continuous, with a bias towards the early parts of a course. The performance of the class as a whole can also inform the teacher of class weaknesses and strengths and so provide some overall feedback.

The teacher needs to encourage self-evaluation and provide resources or strategies to aid this. Providing self-tests that students can do repeatedly to check their improvement in understanding can be of use here.

General questions presented to the whole class to provide feedback to the teachers are often done in an ad hoc fashion, on the spur of the moment. However, it is better to plan this activity carefully. One consideration is to ask questions that deliberately probe different levels of understanding.

Whatever technique is used there is a need to encourage responsible self-assessment by students. There needs to be a parallel commitment to the value of formative assessment by the teacher, a sharing of learning outcomes and a confidence that all students have the ability to develop and improve in statistics.

1.5 Feedback

The two directions of feedback are from the lecturer to the students and from the students to the lecturer. Both are necessary for effective learning and teaching. The students need to know how they are progressing and lecturers need to know the effectiveness of their teaching and of the students' learning. For each of these there is general feedback about overall issues of class progress and progress through the curriculum. There is also specific feedback on the individual student's understanding of individual topics and the teacher's approach to them. Feedback, which can apply to the teacher as much as to the student, has been shown to be a key element in the learning process. Snelgar and Maguire in Chapter 3 discuss the use of feedback on very early assessments in a staged approach to quantitative projects.

Formative assessments of stages in a course, a central element of the feedback, are often supplemented by the use of general questionnaires to obtain feedback from students. Many universities carry out regular surveys of student opinion and experience, the outcomes of which can give useful broad views but they need to be supplemented at both programme and individual course level. To get further detail, an end-of-course questionnaire is often used. In one course students were asked to rate the following topics on a five-point scale: quality of teaching; formative feedback; handouts; organisation; and opportunity to achieve and contribute. Feedback from such end-of-course questionnaires is useful for planning the next run but it is too late for the students concerned, so it should be supplemented by mid-course feedback. In general it is important that at each level of detail some feedback is obtained before too much learning has taken place. The longer that misunderstandings and misconceptions in statistics are left, the harder they are to deal with.

Possible tools for rapid feedback from the students that are readily available are:

- Short tests to explore understanding from a previous class.
- Minute papers with questions such as, 'What part of today's work was most difficult to understand?'
- Wall sheets, placed near the exit, for students to write comments on as they leave the class.

At the very end of a course it is worth spending somewhat longer with groups of students noting and making suggestions with regard to its good and bad points, the teaching and their operation as a class. To ensure that the feedback is effective and that the teacher has the time to respond, it is useful for the teacher to keep a logbook of comments. This will enable an overall picture to be developed and suitable changes initiated. The logbook should also be used at times of marking to record topics that students have found difficult or easy and to note any problems and misconceptions that become evident.

This topic has so far been discussed in the context of the feedback that is of value in improving the quality of teaching. However, the teacher should also review the appropriateness, structure and order of the topics in the curriculum. Should topic X still be taught in the same way at the same time when the course is next delivered? In a review of the curriculum, one needs feedback from a fresh range of sources. These sources should include:

- Current research.
- General literature on the development of statistics.
- General literature on what topics are finding practical application.
- The work experience of past students.
- Employers.

Statistics has some specific issues that need to be addressed. For example, for statistics within a service, or non-specialist course, the students' perceptions of the link between statistics and their main area of study need careful exploration. Such feedback can help in the choice of illustrative material later in the course.

Feedback to students tends to be at an individual level. It is often helpful to give comments on the overall results of the class – common strengths, weaknesses and misconceptions. This approach is also of value when giving end-of-course feedback to students who will often go on to use their learning, and benefit from encouragement and support at this stage.

Feedback to students on their progress should include:

- The precise knowledge, understanding and skill that they should possess at the current stage of a course.
- The level of knowledge, understanding and skill that they have actually reached, and what the gaps are.
- How to go about closing these gaps.

Feedback frequently tends to focus on the second of these elements. One approach that goes some way towards addressing all three is to start each class with a short, student-marked test on the work covered in the previous class. Anecdotal evidence suggests that the most effective way of improving standards of learning is to improve the quality of feedback provided to students – we propose that more research be conducted into this area.

1.6 Summative assessment

Summative assessment needs to be carried out against the formally stated learning outcomes and assessment criteria of a course. There should be a clear match ensuring that all learning outcomes are covered. This may need an imaginative use of assessment methods. If it is felt that some learning outcome is not and cannot be covered, then the learning outcomes need to be re-specified.

The characteristics of summative evaluation are that:

- It is formal, in the sense of being official and to be taken seriously.
- It gives marks or grades.
- It may have several components, but between them they represent the whole of the course.
- It often contributes to the grade and so requires particular care from the assessor.
- It often occurs towards the end of the course.

There are many ways of carrying out summative assessment, with alternatives to consider within each method.

1.6.1 The examination paper

Traditional examination papers are based on a selection of questions to be done in a fixed time. Issues of importance are the comparability of level of questions in a paper, course coverage that is equitable and fair, and the need to ensure that the weaker students have a chance to pass and that the brighter students are challenged. This can be achieved by using structured questions, with sections that are progressively more demanding. It is then essential to ensure that getting the first part wrong is not a total barrier to progressing further with the question. Assessors and moderators need to look at this aspect of papers and examination boards need to examine the mark distributions in papers to assess issues of fairness. The classic approach has been to ask a question on standard theory at the beginning of a question and then ask for it to be developed or applied in the latter parts. This can lead to an excessive emphasis on 'bookwork' and to the mistake of assuming that such bookwork can be classified as 'easy'. Many applications, developments and scenarios can be either segmented into, or introduced by, a progression from easy to more difficult elements.

In statistical data analysis and inference courses, and also in courses on probability and statistical modelling, questions and even whole papers use the analysis of data as the means of testing understanding of statistical ideas and processes. One type of question that is typical of this approach gives the background to a real problem and a copy of computer output of various analyses, and then asks for comments or a report, depending on the balance within the paper. Such questions can sometimes provide a direct link to the student's specialisation in a service course. As mentioned before, care is needed to avoid complicating the statistics by contexts that are not yet sufficiently familiar to the students. Examiners should avoid making the statistics more difficult by assuming knowledge that the student might not have; cultural and language issues must also be given careful consideration. Another means of making links, or of providing a direct application for a specialist student, is the design of questions round a pre-read research paper. Other forms of question include: essays on general topics; questions requiring the design of surveys or experiments; the construction, fitting or criticism of models; and, for the more mathematical course, the proof of theoretical results.

1.6.2 Using multiple-choice questions

Multiple-choice questions have advantages in that they are relatively quick for students to answer, so the range of ideas tested can be wider than in a traditional examination. They can also be marked more easily, by using a grid or by using computer marking. They take thought and care to set, as the 'wrong', or distracting, answers play an educational role in the testing. Depending on the question, they need to be as plausible as the 'right' answers and/or embody common mistakes. Ideally all distracters should be equally 'wrong'. There is also the limitation of needing questions that have clear and specific answers. However, this 'limitation' is the downside of the necessity for the examiners to think very carefully about what they want the students to know and do, which is a positive benefit of this approach to assessment. A style that can be useful in statistics is to ask students whether particular statements are appropriate or inappropriate. This is useful in assessing students' understanding of such aspects as interpreting plots and graphs or computer output from analyses. An important use of multiple-choice testing is for formative assessment. Here interest is not in the right–wrong decision but in the nature of the misunderstanding that led to a particular wrong answer. An aspect of multiple-choice questions that needs consideration, and is less restrictive than often thought, is the marking scheme. Multiple-choice questions do not have to be equally weighted, provided students know what the 'marks' are.

One great advantage of feedback obtained in this way is that if online computer methods are used the assessments can be taken and marked very quickly. This makes quick regular checks on students' understanding possible for the timely provision of support. A further advantage available in many computer-based testing systems is that each student can be given data randomised for a given model and then marked accordingly. This clearly reduces problems of plagiarism.

It may be noted that studies of computer-based assessment find, on the whole, no significant difference in students' assessed levels between paper-based assessments and their computer-based equivalents. See Thelwall (1999, 2000) and the references therein. In both approaches, what matters is the quality of the question design and of the marking.

1.6.3 Short answer, or comment questions

As well as the commonly used 'tick the correct answer' responses, a multiple-choice question environment can be used to elicit short phrases or comments from students. Such phrases are most efficiently received and assessed via a virtual learning environment. Multiple-choice questions usually involve the student scrutinising statements made by the teacher, whereas it is sometimes more appropriate, and may give better assessment coverage of a course, if a teacher tried to assess how students might respond to a statistical statement in their own words. This approach can be particularly useful in data analysis where a scenario and computer output are provided but, instead of asking for a report under examination conditions, a series of questions requiring only a sentence or phrase response are asked. This goes beyond the limitations of providing answers for multiple-choice questions, without requiring report writing under examination conditions and time restrictions.

1.6.4 The individual assignment

There are several advantages in the use of assignments. They can face students with open-ended problems and with issues of problem definition, and also enable a variety of ideas and methods to be combined and applied in a real context. The role of individual assignments is discussed in detail by other authors later in the book.

1.6.5 The group assignment

An important aspect of the statistician's role is that of being a team member. Therefore the assessment needs to encourage and measure this role on tasks suitable for group work. These include tasks already mentioned as suitable for individual assignments, but involving more complex issues and data sets. To these we can add consultancy for a client, e.g. design of a quality control system for a small local manufacturer, or the mini-project type of activity where there are several sub-tasks to be completed. Classic examples are the survey and the experiment, where there are issues of design, piloting, implementation, data collection, data analysis, reaching practical conclusions, presentation and reporting. Examples of group assignments using car price data and agricultural data are given by Mashhoudy in Chapter 21 and by Lopez *et al.* in Chapter 14.

As with all forms of assessment, the first task of the teacher is to develop clear criteria for the marking of assignments. The next problem is the allocation of marks against criteria. In the case of the group assignment there is, firstly, the breakdown of marks for the sub-tasks, with the flexibility to allow for the fact that different teams may adopt differing approaches, or indeed may be carrying out different studies. Secondly, there is the question of group versus individual marks. Common options are:

- All members of a team get the same mark, which reflects the practicalities of team-work, but may be unfair if one member does not contribute.
- A team mark contributing, say, 80% and an individual mark contributing the remainder. This latter mark may be awarded by the assessor on the basis of some individual task within the assignment or on the basis of interview or observation; alternatively, the mark may be allocated by the team itself, for example by allowing each member to spread a fixed total of marks around the team and then averaging.
- A team mark modified by a multiplicative factor representing the contribution of the individual. See, for example, Goldfinch (1994).

One approach to team reporting and assessment is the poster display project discussed by Griffiths and Sheppard in Chapter 4. Here, the team members, perhaps with the use of one or two large posters, seek to show the problem and their solution in a concise form. This demands clear thinking and is an excellent means of sharing ideas and experience between teams and with a wider group of staff. A set of criteria can be devised for the assessment of the poster display, which produces a team mark to go with individual marks.

1.6.6 Continuous assessment

There are many methods of continuous assessment. Regular written tests provide one method: to relieve pressure and give room for formative assessment, the students can submit a selection of these as a portfolio for the summative assessment. An alternative is to use assignments instead of a final assessment. These may be marked in stages as a form of continuous assessment.

1.6.7 The portfolio

Any course on statistics that emphasises the processes of statistics will seek to assess a wide range of different aspects of the students' learning. This assessment will need to include the student's own constructed knowledge and evidence of their learning skills. The development of a portfolio by the student brings together not just the tasks undertaken during the course but also the student's reflection on them, possibly via some form of journal. It is helpful for the students to be given sample questions to ask of themselves, such as: How did you respond to the feedback obtained from an assessment? How would you improve your work on the task done? What relationships do you see between Topic X and Topic Y? How could the operation of the group on the previous mini-project have been improved?

1.6.8 Self-assessment

Learning theory emphasises the importance of students, both individually and as part of a group, taking responsibility for their own learning. An important part of this is self-assessment. This covers a broad spectrum, from a general sense of progress on a topic to the self-marking of tutorial questions. It also includes students working in pairs or groups to mark each others' work. This type of activity helps in the overall process of self-assessment and is increasingly used by staff, both for educational motives and to help them focus on their supportive role as teachers.

1.7 Techniques and practicalities of assessment

When a teacher is designing an assessment careful consideration should be given to:

- The learning outcomes and the capabilities/skills (implicit or explicit) they imply.
- Methods of assessment that match with these outcomes and skills.
- The relative efficiency of different methods in terms of student time and staff time.
- The advantages and disadvantages of any proposed method.
- The forms of marking scheme or criteria that are appropriate.

This approach should naturally lead to the design of a specific assessment task, but care must be taken to avoid common pitfalls. These include: unintentional ambiguities in a question or assignment, miscalculation of the time and resources required to do the assignment or to mark it, and failure to design either a suitable set of criteria or a marking scheme.

The design of effective assessment tasks in statistics can be time-consuming. Essay questions are quick to set but lengthy and subjective to mark; to be reliable, very detailed marking schemes are required. Mathematical statistics questions often take a long time to set but are fairly quick and objective to mark. Statistics questions in general can be lengthy to set and lengthy to mark, and need careful marking schemes.

It can be instructive to consider questions set by colleagues in other subject disciplines. There will almost certainly be questions that are intriguing and surprising, and also forms of question that may be useful for assessing statistics. Occasionally the content, as well as the form of question, can be relevant to statistics (Brown, 2001).

In considering the assessment of different types of activity, great care needs to be given to defining the process for allocating marks. It is sometimes helpful to have two sets of criteria, one dealing with the specifics of the activity, the other setting out general descriptive criteria for deciding what ranges of marks are appropriate. The marking seeks to reach consistent answers for the two sets of criteria.

Before discussing any methods in detail it is important to underline the role of assessment. Many students view assessment as merely a test of memory after having done all the work. However, in the areas discussed here assessment is an integral part of the learning experience. It ensures that the students carry out the processes of statistics and is thus essentially formative as well as being summative. This fact indicates that more attention needs to be given to feedback to students, even where the assessment provides summative results.

A vital element of statistical assessment is the assessment of students doing statistics. Some of the choices facing the teacher are illustrated below.

1.7.1 Tutorial exercises and examples

The classic student experience of doing statistics is that of doing tutorial problems. A common practice is to ensure that students can start on the problems before the tutorial so that some tutorial time can be dedicated as necessary to clearing up difficulties students identify and discussing the harder problems. Making best use of tutorial time in this way means that well-structured tutorial sheets need to be available to students at least one week before the tutorial session. An informal record-keeping system will enable this approach to be used more effectively for formative assessment.

1.7.2 Data studies

Most students will use their statistics to investigate and analyse data. Thus, most courses will involve studies of sets of data. To be effective the study needs to be well focused. Most data can lead into a range of different lines of study. Students need to be prevented from spending time on peripheral issues and so they should be assisted in understanding some of the issues of choice between alternative lines of enquiry.

For service courses data drawn from the main discipline is clearly a good source for these students and will also interest them. The complexity of the study will depend on the tools available for any analysis, in particular the software available.

1.7.3 Practical work in the classroom

Rather than being supplied with data, it can be a valuable experience for students to collect the data for analysis themselves, in specified classroom experiments. The potential pitfalls in conducting these types of experiments have led to such practical work rarely being used for summative assessment. This is not a major problem in formative assessment, for which such practical work is commonly used, though time pressures for staff and students seem to have reduced the use of practical work actually carried out in a classroom environment.

1.7.4 Investigations

The availability of data on the Internet provides a vast resource for students to carry out investigations into problems of relevance to their main study or of current interest. Ideally an investigation will be centred on a real problem. In such investigations student assessment will probably be based on three elements:

- Ability to use statistical knowledge.
- Strategies adopted to solve the problem.
- Ability to communicate and report on the problem and its solution to the problem owners.

1.7.5 Consultancy

For postgraduate study, many US universities have an 'intern programme'. This provides a university-wide consulting service using staff and students. Some UK universities have used student teams to provide consultancy to local businesses and industries, or within the university. Assessment may be based on both written and verbal interactions with, and presentations to, the client. Consultancies can also be simulated, with a staff member – preferably from another department – acting as the client and the work centring on a scenario with data/information revealed gradually, as the student asks for it.

1.7.6 Case studies

Case studies provide an application-oriented form for teaching. The assessment tends to take the form of reports on the case to the 'problem owners'. This promotes key problem-solving skills, but can miss the assessment of detailed technical knowledge. One way of tackling this is to generate artificial and specific problems that test knowledge, but which are set within the context of the case study. As usual it is essential that students have detailed information on the marking methodology of all assessments within the case study.

1.7.7 Mini-projects

Mini-projects can be built around finding information on a specific topic or methodology. They can also be part of a major project, such as fitting a defined model to a set of data from the major project. The assessment here would follow that used in a major project. Due to the mini-nature of the work, the assessment would tend to be of the final output rather than of the process.

1.7.8 Projects with dissertations

For final-year undergraduate and MSc projects, the assessment tends to be based on the final output, but the time allocated also allows some consideration of the assessment of the process of the research. This may require a detailed research strategy and project plans after the first month of work or take the form of midway presentations on work in progress or course poster sessions on the research in the latter stages.

1.7.9 Statistical problem-solving

Unfortunately, as Marriott *et al.* (2009) show, assessing statistical problem-solving needs far more thought than at first meets the eye. They demonstrate that teaching and learning statistical problem-solving requires a paradigm shift in teaching and that the cognitive skills expected of students are not what are traditionally expected. It follows that assessing statistical problem-solving is a much more complex procedure than is achieved by assessing through traditional examinations. More research is needed in this area.

With so many approaches, the teacher is clearly faced with frequent problems of choice. One guide in this choice can be found in Anderson and Krathwohl (2001) who provide a taxonomy for teaching, learning and assessment. The authors consider the interaction between categories of what they refer to as the cognitive and knowledge dimensions. An important aspect of their approach is the careful examination of the verbs used in stated teaching and learning outcomes, and how they relate to categories of the cognitive process. Their Table 5.1 provides the detail and a helpful list of verbs.

Anderson and Krathwohl use a table with rows and columns that are the categories of the knowledge dimension and the cognitive process dimension respectively, to show how a teacher might approach the assessment of learning outcomes. In order to illustrate this, consider the learning outcome (LO): 'The students should learn to use confidence intervals for inference'.

In Table 1.1 we provide a taxonomy for this LO.

Table 1.1 A taxonomy table for the assessment of confidence intervals.

Knowledge Dimension		Cognitive Process Dimension → 1 Remember	2 Understand	3 Apply	4 Analyse	5 Evaluate	6 Create
↓	A Factual knowledge						
	B Conceptual knowledge		TL	LO TL	TL	TL	
	C Procedural knowledge	EX		LO TL EX		TL	
	D Metacognitive knowledge	TL		TL			

For the above LO, the verb 'use' means 'implementing' in this context, which is associated with 'Apply' in Anderson and Krathwohl's Table 5.1. In order to accomplish the ability to 'use' successfully, the students must identify the type of problem they are faced with, decide on and select the inferential procedure that is required (the appropriate confidence interval), and then apply the appropriate procedure to compute the interval. If we now look at the verbs that have been used to describe the tasks the students must undertake (identify, decide, select and apply) they would indicate that the implementing in this example requires both 'Conceptual knowledge' and 'Procedural knowledge'. Cells B3 and C3 in Table 1.1 are therefore associated with the learning outcome in this case and are marked with 'LO'. In teaching to this outcome the student would be encouraged to use 'Metacognitive knowledge' by being taught how to assess how they know the problem fits into a particular type. Then they would ask themselves whether the answers they arrive at are 'sensible' at the end of the process of constructing a confidence interval. The students are also taught to 'Evaluate' their answer in context. The cells B2, B3, B4, B5, C3, C5, D1 and D3, marked 'TL' in Table 1.1, are all associated with the teaching and learning needed to support this specific learning outcome.

In planning an assessment, the teacher must now choose appropriate assessment(s). A formative assessment may well be appropriate if the knowledge and cognitive processes identified in the consideration of the teaching and learning supporting the specific outcome are to be addressed. If a strict assessment of the learning outcome is required, the assessment could be tailored to assessing the types of knowledge and cognitive processes associated with cells B3 and C3 alone, for example: 'Calculate the 95% confidence limits for the large sample confidence interval given your sample data'; to answer this question, the student must 'Remember' and 'Apply Procedural knowledge'. (The associated cells in Table 1.1 are marked 'EX'.) It can be seen that this question only aligns with one of the cells identified as applying to the stated learning outcome.

1.8 Assessment of active learning approaches

Individual assessments are supported by detailed marking schemes leading to percentage marks or grades. These are often supplemented by descriptive assessment criteria that link the student's overall performance on the item of assessment to the more general learning outcomes. The two should provide consistent appraisals. Any inconsistency needs careful appraisal by teachers and Examination Boards. Often, two staff will mark a project and agree an overall mark.

Table 1.2 presents an illustrative set of words suitable for assessment criteria for different types of learning outcomes. Three categories of Fail, Pass and Distinction are used. Some criteria seek to subdivide pass into 40–49, 50–59 and 60–69. The detailed marking scheme should normally do this.

One example of an approach to designing a marking scheme and then allocating marks for assessment is provided by Marriott *et al.* (2009). They adapt the delightfully simple grading scheme for UK Advanced Level coursework in statistics provided by the curriculum development organisation Mathematics in Education and Industry (MEI). The scheme allocates the assessment questions to domains for grading and uses a very simple mark allocation scheme suggested by Garfield (1994, Example 2). Allocated marks correspond to the responses candidates make to each question being incorrect (0 marks), partially correct (1 mark) or correct (2 marks).

1.9 Conclusions: assessment strategy – principles and guidelines

In planning for success in assessment, it is useful to take on board some core principles and follow succinct guidelines.

- *Principle 1:* Content – assessment should reflect the course content that is the most important for students to learn.

Table 1.2 Assessment criteria for different learning outcomes.

Form of Learning Outcome	Fail	Pass	Distinction
Knowledge of the theory of least squares.	Shows no knowledge of theory of least squares.	Appropriate range of knowledge of theory of least squares.	Theory of least squares used critically and innovatively.
Comprehension: Understand the method of least squares as applied to linear regression.	Significant lack of understanding, misunderstanding of both method and regression.	Adequate to competent comprehension of least squares and of its application.	Explanation comprehensive, accurate and insightful.
Application of the method of least squares.	Poor and erroneous use of method.	Largely correct and appropriate use of method.	Accurate and insightful use of method.
Analysis of a real problem using a data set.	Few ideas and lack of recognition of context.	A reasoned set of analyses set in the problem context.	A comprehensive and insightful set of analyses, clear conclusions for the context.
Synthesis (i) Carry out a library-based project on a new area of application and produce a report in survey form.	Little apparent understanding of the area and an unstructured and confused report.	Good review and coverage with clear, balanced and structured explanation. Well referenced.	An insightful and critical report, Excellent coverage of the literature with an integrated and concise structuring.
Synthesis (ii) Design and carry out a small survey.	Design inappropriate to objectives. Poorly implemented.	An appropriate design competently carried through.	Design and implementation show depth of insight and thoroughness of implementation.
Evaluate a published study involving the statistical analysis of medical data.	Inadequate appreciation of the problem, the data and the analysis.	Correct understanding of the problem and methods, with a correct identification of the strengths and weaknesses of the methods used.	Shows critical insights into the issues and methods involved with the correct and imaginative advocacy of possible alternatives.

- *Principle 2:* Process – assessment should enhance the sound development of concepts, insights and deeper ways of thinking; it should encourage deep learning.

- *Guideline 1:* Students value what is tested, so test what you value.
- *Guideline 2:* Doing formative assessment implies that the student wants to know as well as wanting to pass.
- *Guideline 3:* Good summative assessments will seek to ensure that the student understands as well as passes and that all passing students have reached a basic competency.
- *Guideline 4:* If formative and summative assessments do not match then generally the summative will dominate, as students are generally more interested in passing than in understanding.

These six principles and guidelines require that the stated LOs and assessment criteria clearly reflect what is valued by the students, are relevant, and are important for students to understand.

A final consideration in deciding on an assessment strategy is the sheer practicality of carrying out the marking involved. Consistency of marking is important, and one way of ensuring this is to have more than one person marking each item of assessment. It is also important to remove any possibility of personal prejudice in marking; one way of doing this to use anonymous marking. It is possible to develop efficient techniques that fit the statistical content, the type and range of students and the requirements of the learning outcomes. Some of these often raise concerns as being too time-consuming. It should be noted that, with experience and ingenuity, most forms of assessment can be carried out effectively and efficiently. In essence the teacher's best assessment strategy is to choose the best balance of methods to achieve the given ends.

2

Variety in assessment for learning statistics

Helen MacGillivray

2.1 Introduction

Assessment is for learning. Awareness of the importance and implications of this statement is escalating. The general higher education literature increasingly emphasises the role of assessment in learning (Angelo, 1999), and of the explicit aligning of assessment with outcomes (James *et al.*, 2002). There are now more rich and varied possibilities in learning and assessment strategies available than in previous eras, and better understanding of the roles of assessment in learning. Within the discipline of statistics there have been similar calls for statistics educators to assess what they value (Chance, 2002), and to meet the assessment challenge in a field that is fundamentally vital wherever there is quantitative information, variation or uncertainty, but in which it is notoriously difficult to facilitate understanding and the learning of its concepts and ways of thinking.

Research in learning and teaching in statistics, considerations of the nature of statistical enquiry and how statisticians think (Wild and Pfannkuch, 1999), and general educational research are combining to develop principles, strategies and resources in statistics teaching in higher education. Teacher-centred strategies, with theory followed by examples that often tend to be isolated and context-poor, are being supplanted by student-centred, data- and context-driven, experiential learning, with emphasis on concepts and the development of statistical thinking. However, the richness and universality of statistics in real contexts and data contribute to the challenges of designing assessment for learning statistical

Assessment Methods in Statistical Education Edited by Penelope Bidgood, Neville Hunt and Flavia Jolliffe

thinking, particularly in catering for diversity in student cohorts and in educational cultures and attitudes. Designing learning and assessment packages in statistics requires choices and balance, and hence understanding of the variety of assessments possible and their roles in contributing to a holistic approach.

After a brief overview of the roles of assessment in learning, this chapter presents arguments for the importance of variety in statistics assessment from three perspectives, and considers some general principles of criteria and standards in assessment and creating learning environments in statistics. It then discusses a wide range of types of assessment in statistics and their roles in student development, including their capacity for variety in contexts, data and problem-solving. Where appropriate, strengths and weaknesses of various types of assessment are discussed, including their potential to contribute to the full diversity of statistical thinking. The aim is not to give details of specific statistical topics and specific examples of assessment items, but to provide insight and assistance for choosing and developing types of assessment components within the designing of a balanced and integrated learning and assessment package appropriate for a course and cohort.

2.2 Assessment is for learning

This statement has many dimensions to it. Assessment is for learning by the student – of knowledge, skills and thinking in a field. The universal importance and the nature of the discipline of statistics are richly and endlessly challenging for the creation, implementation and management of authentic assessment for developing student statistical thinking, skills and confidence, whether foundational or more advanced.

Assessment is for learning about students and their work – by teachers and by students about themselves. It is through assessment that students learn about their individual strengths, weaknesses and current level of skills and understanding. Assessment that helps students to develop understanding of themselves as individual learners is clearly of prime importance. Such assessment also facilitates active engagement by students, and helps them understand and value their assessment tasks (Angelo, 1999). Assessment is also for learning about students by the instructor(s), not just for certification, but also to feed back into the teaching of the current cohort and for the future. Starkings (1997: 1) describes a key role of assessment as the 'diagnostic process – by establishing what students have learned, it is possible to plan what students need to learn in the future'. Although this statement refers mostly to planning for a current cohort, it also applies to the ongoing development of teaching for future cohorts.

Assessment is also for the learning of generic skills or graduate capabilities, which are now often explicitly articulated and emphasised, particularly in university education. Whilst it is important that these are identified and integrated within and across the curriculum, they cannot be learnt in isolation – they are what one learns while learning something else. The learning of statistical literacy and thinking is a rich environment for the development of the generic skills viewed as desirable in many expressions of graduate capabilities. For example, Petersen (2005) comments how statistics has been used to teach writing for many years in a CHANCE course.

Thus, assessment is a living and active process that produces, develops, moulds and diagnoses learning for students and teachers, and (Wild *et al.*, 1997) is 'the most powerful signal we have for telling students what we believe to be important'.

2.3 Roles of variety in assessment in statistics

There is increasing emphasis (Pfannkuch and Wild, 2004) on variation as the core of statistical thinking, and the integration of the statistical with the contextual. Awareness and understanding of the messiness of data, the importance of context and the effects of assumptions in interpretations of statistical analyses and models, are key aspects of statistical thinking. To learn how to think about, interpret and handle variation, students need to experience it. The greater the variety of authentic data sets and contexts with which students work, the more confidence they develop in statistical thinking. This variety should be an impetus in designing assessment, whether formative or summative, for learning. This is the first aspect of the need for variety in statistics assessment.

Across school and introductory university levels, much of the emphasis on data and context in statistics education literature has been on data analysis. However, the above comments apply to levels beyond introductory university and also to statistical modelling in its broad sense of including stochastic modelling. Statistical modelling is the modelling of any situation that involves variation and uncertainty. Integral to any such situation are context, data, probability and distributions; it is the nature of the particular context and the problem under consideration that determines the composition of the mix. To advocate that probability is merely a branch of mathematics is to risk trivialising statistics and statistical thinking. The implications of this for stochastic modelling are at least twofold: it must also be driven and built around authentic contexts, and it must link with data. Perhaps because fewer students are involved in such courses and because of society's imperative needs for statistical literacy, university courses in statistical modelling (in its broad sense) and in data analysis beyond the introductory, have featured less in statistical education literature. This does not mean that innovations in teaching post-introductory levels and stochastic modelling are not happening, but more attention is needed in these areas.

Investigating and understanding data in context with associated nuances of interpretation is a cornerstone of statistics, but statistics is also a systematic, quantitative and coherent discipline with concepts, principles and procedures. It has a balance of rules and freedoms, of details and synthesis, of correctness and subtleties. Indeed, although interpretation in context is a critical component, a major element of the power of statistical thinking lies in its transferability across contexts, problems and applications. The scope of statistics includes: operational knowledge and understanding of statistical principles; choice and use of statistical procedures; planning and conduct of data investigations; data handling and processing; interpretation that combines statistical principles with contextual distinctions; understanding and evaluating assumptions of models; choice and application of models; and choice and application of mathematical tools and thinking. Facilitating learning across even a subset of such a range requires carefully designed programmes with a variety of types of assessment. An assessment type that is appropriate for learning one aspect may be less appropriate for another. Thus, the second aspect of the need for variety in statistics assessment is variety of types of assessment.

The third aspect is common to all disciplines. This is variety in assessment to cater for different ways of learning (Felder and Spurlin, 2005), such as sensing or intuitive, visual or verbal, active or reflective, sequential or global. Although details of these are not pursued, much of the discussion here of the diversity of statistical thinking and of the associated variety in assessment will include catering for different ways of learning. Awareness of this general need for variety can contribute to the overall richness of a learning and assessment package.

Awareness of the importance of assessment for learning and of the roles of variety in assessment in statistics is demonstrated in writings about teaching statistics in general, and not just in papers focused on assessment. For example, in every contribution in Garfield (2005), assessment is quickly introduced and emphasised, with variety in assessment a common thread. All contributors in Garfield (2005) are from one country, but their comments are varied. For example, Evans (2005) lists 14 types of assessment used in one course. Although these are described as being to measure student learning, it is clear they all have their roles to play in a carefully designed course plan and timeline of learning.

2.4 Statistics within some general principles of assessment

Gal and Garfield (1997a: 5) wrote 'The unique challenges that statistics teachers face stem from the existence of multiple subgoals . . . which require teachers to address a wide range of conceptually distinct issues during instruction. Educators are further challenged by the need to make sure that students understand the real-world problems that motivate statistical work and investigations'. In a survey of statistics educators in the United States, Garfield *et al.* (2002) reported that of all areas of statistics education, assessment practices had undergone the least reform. The many dimensions of the assessment challenge are complicated in introductory courses by the diversity of student cohorts in which the wide range of backgrounds, programmes, motivations and study skills need consideration in designing appropriate assessment and learning packages. Moreover, the effects of educational cultures and histories of different countries that are reflected in student and staff expectations and habits are more evident at introductory university levels.

The assessment challenges described in Gal and Garfield (1997a) are associated with the complex nature of statistics and its learning, and with the needs for variety as discussed above. But implementing variety in assessment requires balance and constructive alignment (Biggs, 1999a) to achieve a harmonious progression of learning development. Biggs (1999b) describes this as a fully criterion-referenced system, in which all components of the system address the same agenda and support each other. This incorporates Rumsey's (2005: 85) call for seamlessness, in a 'big ideas and common threads' approach.

There is considerable misunderstanding about criterion-referenced assessment, with many believing it requires verbal descriptors of criteria and standards in which numerical scales have no place in representing levels of achievement. In contrast, O'Donovan *et al.* (2004) conclude that over-reliance on explicit descriptors is as naive as past over-reliance on tacit knowledge. A characteristic of true criterion-referenced assessment is transparency. Good learning and assessment packages are integrated, balanced, developmental, purposeful packages with well-structured facilitation of student learning across the cohort diversity. Such packages possess an inbuilt configuration or pattern of performance as required by Sadler (1987) for standards referencing. The configuration comes about through a combination of the construct of formative and summative assessment (aligned with student learning across the spectrum appropriate for the purpose and cohort), and the construct of timing, types and weights of assessment tasks. Sadler emphasises the importance of exemplars (such as marked past student work, representative assessment tasks and model solutions) in identifying the characteristics (or criteria) of each component of assessment, with verbal descriptors to draw attention to salient criteria at different points. Indeed, interviews with students indicate that they tend to ask for exemplars and verbal descriptors of criteria for holistic assessment such as open-ended projects and reports, but request only exemplars for assessment such as tests or assignments emphasising knowledge and procedures. Students value exemplars for all tasks, including information on relationships of previous tasks or student work to the current learning situation. Students tend to regard 'ticks in criteria boxes' as poor feedback, and to condemn a combination of grades over components of assessment without an explicit weighting schema.

Debates about marks versus textual grades are frequently misleading. It is the representation of criteria and standards, and the quality of feedback to students that are important. Ongoing feedback to students is essential to learning, but it must be feedback that helps them identify their strengths and weaknesses and provides practical, efficient and effective guidance on how they can progress. It is not surprising that different forms of marking and feedback may be appropriate for different types of assessment. The overriding and unarguable requirements are that marking or grading of assessment be consistent, transparent and accountable. As standards referencing comes from the construct of timing, types and weights of assessment tasks, and an explicit weighting schema is essential for transparency of overall grading, particularly at university level, quantitative representation of levels of achievement is required at some stage, and hence should be explicit.

Overuse or misuse of verbal descriptors and coarse grading schemas tend to obscure rather than enlighten.

2.5 Creating learning environments in statistics

Biggs (1999a: 61) writes: 'Learning is the result of learning-focused activities which are engaged by students as a result both of their own perceptions and inputs and the total teaching context'. In Biggs' constructive alignment, students are 'entrapped' in the 'web of consistency, optimising the likelihood that they will engage the appropriate learning activities' (Biggs, 1999a: 64).

Both general higher education literature and statistics education literature emphasise the importance of active learning and student engagement. In statistics (and mathematics), passive learning is no learning; the learning comes through doing. The operational knowledge and skills of such areas must become more than just familiar to a learner – they need to become part of the learner. Helping students acquire the confidence, skills, thinking and useful dispositions (Gal and Garfield, 1997a) of statistics requires many interwoven components in a teaching, learning and assessment package. Variety in statistical experiences and in types of assessment is an essential component in creating learning environments which help students own their learning and sustain their engagement.

Evans (2005: 71) speaks of 'authentically shared learning experiences' and Chance (2005: 101) of it being 'much more fun as the Teaching Assistant than the teacher'. Such comments reflect the benefits of students working 'collaboratively and in dialogue with others, both peers and teachers' (Biggs, 1999a: 61). Research is indicating that student–teacher collegiality is important for student learning in disciplines such as statistics and mathematics (Solomon *et al.*, 2008). Constructing environments to facilitate this is not restricted to small classes. Although it may take considerable thought, experience and planning to achieve staff collegiality with students in a large class, the key is in the attitude and culture of sharing the journey. Far from diminishing the teacher's standing and authority, taking the role of coach and team-mate tends to increase students' respect, provided the overall learning and assessment package is a well-structured program with clarity, purpose and harmony in content, objectives and aligned assessment.

Much is written about active learning as a problem-solving process, but the creation of an environment for learning problem-solving is a significant challenge. Constructing assessment that assists in creating such an environment depends on the course objectives, content, cohort and level, but variety of authentic statistical contexts and different types of assessment can provide a range of opportunities. In contrast to Biggs' (1999a) definition of good and poor learning habits, stronger learners in statistics tend to first explore, question and experiment in the learning environment designed for them. If learning experiences are inadequate in variety, and shallow in authenticity and depth, then students may feel the need to go beyond the immediate learning environment provided to them. But the diversity and balance of concepts, systems, procedures, interpretations and nuances in statistics, can cause confusion for students if they try to experiment in more than one learning environment simultaneously. The need for well-constructed environments with lots of scope for learning is driven by the natural variety in authentic statistics contexts and data, and can be met through variety within and across assessment tasks and types.

2.6 Assessment for learning through discovery

All assessment is guided and controlled at least to some extent. Indeed, it must be in order to meet learning outcomes and enable identification and application of criteria and standards. The first types of assessment considered here are the more open-ended types. They have the advantages of student discovery, ownership and synthesis, but the disadvantages of focus on contexts and dependence on student understanding of their current statistical knowledge and skills. Students need either to have already acquired an appropriate set of statistical skills, no matter how basic, or be able to acquire these and appropriate statistical dispositions during their learning and assessment progress. The aspects of statistics emphasised are: those of choice and use of statistical procedures; planning and conduct of data investigations; data handling and processing; interpretation of statistical analyses in context; understanding and evaluating assumptions of models; and choice and application of models.

2.6.1 Full data investigations in a free-choice environment

A very open-ended form is the data investigation project in which students identify their own topic to be investigated, plan and implement a data collection strategy to investigate the topic, explore and analyse their data, and produce a written report. This is usually a group project because the task needs a group in the brainstorming, the planning, the data collection and the interpretations of results in context.

Such a process is at the heart of statistical thinking in empirical enquiry (Wild and Pfannkuch, 1999) and reflects the Plan-Do-Check-Act cycle (Shewhart and Deming, 1986) the Problem, Plan, Data, Analysis, Conclusions model of MacKay and Oldfield (1994) and the Data Handling cycle of Plan, Collect, Process, Discuss (Marriott *et al.*, 2009). These approaches to data investigations are appropriate across all levels of education, with opportunities for demonstration of increasing maturity, thinking and skills in all parts of the cycle.

A significant impetus for learning is student ownership – of the ideas, the data and therefore the analysis. This is emphasised in MacGillivray (1998, 2002) in which the benefits, challenges, strategies and practicalities of this approach in large classes are also discussed. Chance (2005:105) describes such projects as 'the most valuable learning experiences for my introductory students'. Lee (2005) describes changing from projects on a real-world problem posed by the teacher to projects on topics chosen by the students, commenting that students tended to be uninterested in working on real-world problems chosen for them, with analysis and reports on their chosen topics tending to be better.

At any level, the objective of this type of assessment is the experiential learning of the whole process of statistical enquiry – of the challenges of turning ideas and questions into plans for investigation, of the practicalities and messiness of data collection and handling, of the essentials of choosing and using statistical tools, and of the synthesis of statistical interpretations in real and authentic contexts. To fulfil this potential, the topic and collected data should contain a number of variables, preferably including both quantitative and qualitative data. Possible weaknesses include context domination or over-immersion in context, excessive enthusiasm or ambition in topic choice, and management of teamwork. Experiments, observational studies, surveys, workplace data and secondary data sources all have potential strengths and weaknesses. Ongoing guidance and support for students and teaching assistants can include exemplars of model projects, marked past projects, criteria and standards, and overall clear identification of the learning outcomes of the project and its role within the learning and assessment package.

Criticisms of the above free-choice discovery projects at the introductory university level tend to be of three types. The first acknowledges that such projects provide invaluable experience, but are time-consuming for students and staff. The second expresses concerns that, at the introductory level, such projects may not be conducted 'correctly' by students; and the third, that the topics investigated are not sufficiently 'meaningful'. The first point is a valid concern, and care must be taken in constructing and managing student and staff support, and the whole learning and assessment package, if this type of assessment is included, particularly in large classes. For example, because this type of assessment is holistic and focused on the overall development of statistical thinking, other types of assessment can focus on operational knowledge and understanding, and be of types that are less time-consuming and more easily regulated in marking. The second criticism tends to overlook the potential for such projects in learning to question in statistical thinking, whether it is questioning of the data quality, of assumptions, of choice of procedures or of interpretation of output. The third type of criticism tends to be indicative of beliefs that students will appreciate and understand statistics only if they see it used in their own disciplines, or in 'important' contexts. To learn how to think about, interpret and handle variation, students need to experience it. Unfamiliar or complex contexts can inhibit statistical learning. A context chosen by students for their statistical learning through discovery is meaningful for them.

2.6.2 Smaller data investigations in a free-choice environment

At any level, the extent of statistical skills available to the students – whether already acquired or being acquired in parallel with the investigation – provides the parameters of expectations of the students as well as guidance for them. Smaller investigations can target specific statistical procedures while retaining the elements of student choice of variables to investigate, the practicalities of planning and collecting data, using and interpreting statistical procedures in context. An advantage is the more incremental approach. Potential weaknesses are associated with domination of procedure – selecting example or data to fit a procedure – and simplistic single answers. Strategies for avoiding or lessening these weaknesses include raising questions or suggestions at each stage of the process – what to investigate, assumptions of procedure, relevance of interpretation and possible future investigations.

2.6.3 Data investigations in modelling with probability and distributions

Moore's (1997) observation that students need only an informal grasp of probability to follow the basic reasoning of statistical data analysis, is both correct and astute, and any probability considered in an introductory university course whose focus is data analysis should be purposeful and minimal. If an introductory statistical course is required to include probability and/or introductory distributions, the purpose, placement and structure should be clear, both within the course objectives and within the students' overall programmes.

However, probability, distributions and modelling with them are the heart of statistics, and are of increasing importance for students in other disciplines as well as for future statisticians. Whenever and wherever they feature in students' programmes, they need as much thought and reform as inference and data analysis. Early and ongoing linking with data helps to integrate and develop the concepts of chance and its models that underpin all of statistics.

Early links with data can vary from estimating probabilities in conditional situations, in simple Markov chains and in distributions. Probably the most common reaction to linking data and stochastic modelling would be time series, but there are many other possibilities in less sophisticated contexts. For example, data investigations in free-choice environments are readily accessible at relatively early stages in stochastic modelling in the context of queues and other possible Poisson processes (MacGillivray, 2007). The great variety of everyday contexts available, the practical questions and challenges of collecting data from either the number of events or the time between events, and the combination of using goodness-of-fit and graphical procedures for assessing Poisson assumptions, make this a fertile discovery environment for learning and assessment.

2.6.4 Simulated environments for data investigations

Simulation power and possibilities have opened as many exciting vistas in statistics education as in statistics research. The breadth and depth of recent innovative work in simulation systems for teaching statistics are providing rich and complex environments for learning and assessing statistical thinking and authentic challenges in statistical analysis beyond the introductory university level. The virtual experiments of Darius *et al.* (2007), the interactive documents of Nolan and Lang (2007), and the problem-solving simulation environment (called Watfactory) of

Steiner and MacKay (2009), all provide rich, holistic environments for more advanced students for authentic assessment in meaningful learning through discovery. These systems are highly appropriate in developing a set of advanced statistical skills, as the discovery aspect in these requires prior familiarity and confidence with at least some statistical analysis tools. But the power of each of the above systems in learning and assessment in statistics lies in variety of experiential learning – the variety of data, of choice of analysis procedures, and of interpretations in context of the output of statistical analyses.

2.7 Assessment and learning in given scenarios

Familiarity, understanding and confidence with knowledge and procedures in statistics come through exercises that manage learning in small components of the development and progression. Variety of contexts and data provide the experiential learning and the formative assessment, and the exemplars for summative assessment.

2.7.1 Exercises and practical activities with data

Whether on paper or in computer laboratories, exercises that provide a range and variety of data experiences are essential for statistical learning and assessment that develops and evaluates statistical skills and understanding. The value of real data and real scenarios is now generally accepted, but small data sets and isolated scenarios seldom occur in real situations. Assessments that focus on parts of the data investigation cycle and that clearly demonstrate these parts within the whole process are better reflections of reality and help fit the incremental components into the full statistical story. For example, using a complex data set and asking students to extract what is needed to answer questions is just one important step in developing statistical thinking. Other aspects of the data investigation cycle that could be considered in assessment for learning include: critiquing the plan and data; proposing questions to be investigated; planning a new investigation; and conducting a pilot.

2.7.2 Exercises in probability and distributional modelling

As with the facets of statistics primarily concerned with data investigation and analysis, variety and authenticity of contexts in stochastic modelling are essential for real learning, and, again, should be cornerstones in assessment. In these statistical areas, constructivism, creating problem-solving environments and building problem-solving confidence, are all of great importance. Variety of contexts that incorporate the everyday, familiar or authentic, contributes significantly to student development (MacGillivray, 2008). Chance plays a large part in our everyday and life decisions, yet formalising and using concepts of probability and distributions can be challenging. It is in such situations that formative assessment of prior learning can assist in consolidation and extension – colloquially described by one student as 'using what we already know to learn other stuff is really good and helps us learn other stuff'.

These are also the areas in which the emphasis on mathematics as separate to, but good servant of, statistics can be of great value to students' development in both mathematical and statistical thinking. For example, McColl *et al.* (2007) present an online environment that separates the contexts of distributional models from their mathematics; this can be used in either formative or summative assessment. At more advanced levels, separation of the modelling with probability, the serving mathematics, and applications of models, becomes even more important. Formative assessment that links with prior learning and identifies the three components noted above assists development of skills and confidence.

Designing types of assessment other than given scenario exercises and problems is more challenging in these areas than in data analysis. Some aspects of linking with data and data investigations are mentioned in Section 2.6.3. Participation assessment is discussed below in Section 2.8. At levels beyond introductory, researching the why, how, when and who of developments can assist student understanding as well as adding to the variety in types of assessment. In both data analysis and stochastic modelling, the people and history of developments can be strong motivators for students to understand the statistics.

2.7.3 Case studies

Case studies are highly effective in exposition, motivation and demonstrations, illustrating statistical thinking and the power of statistical methods. In assessment, their focus is necessarily on analysis. It is possible to include something of their role within the full data investigation process, but the development of topics to be investigated, plans and their implementation, are usually fixed or at least partially fixed by the case study. Full background information is needed to enable aspects of the data investigation cycle other than analysis and limited interpretation to be included. One of the key difficulties with case studies as vehicles for experiential learning and assessment, particularly at the introductory level, is the context. Unless the context is already familiar or easily grasped, the student has to learn about the context as much as, if not more than, the statistical concepts and thinking. This can inhibit or hide the statistical learning, and reduce its transferability. In using case studies for assessment, care is needed to avoid the assessment being more about the context than about the statistics. Case studies with rich and complex data tend to be more powerful vehicles for experiential learning and assessment at advanced statistical levels in training for the statistical profession.

2.8 Attendance, participation and assessment

Aliaga (2005: 76) comments on students' having 'moments of readiness', and on their 'need to talk, debate, disagree and argue'. Many statistical educators emphasise active learning, but comment that students need encouragement to participate and share their learning experiences with their peers and teachers. One of the most common challenges for university staff is the bimodality in attitudes to participation amongst university students. Although there appear to be more conflicting demands on students, such as paid work, than in previous eras, some common reasons given by students reluctant to attend or participate include their feelings of needing to 'catch up' by themselves, and of not exposing themselves to any form of assessment until they feel prepared. One of the most puzzling aspects is that very few students tend to miss tests, no matter what the assessment weight, but miss assessment items such as assignments or practical reports for which considerable assistance is available.

Thus, some students need more encouragement than others to take the step into participation. A clue to this apparent dichotomy may be found in Gal *et al.* (1997b: 2), who describe an ideal environment for learning problem-solving as 'emotionally and cognitively supportive' and one in which students 'feel safe to explore, feel comfortable with temporary confusion, believe in their ability to navigate temporary roadblocks, and are motivated to struggle and keep working'. One way to encourage all students into such an environment is to allocate a small amount of assessment weight to active participation in the type of environment described above, where all the emphasis is on the act of active and collaborative learning. When it is remembered that credit in assessment is for achievement in learning, it can be seen that inclusion of this type of assessment is fully justifiable in an overall balanced design (MacGillivray, 2007).

2.9 Quizzes, multiple-choice questions, tests and examinations

Quizzes, tests and examinations play far greater roles in overall learning and development than they are often credited with. Understanding, choice, application and interpretation of procedures can all be assessed in test conditions through carefully designed stages of short responses. Assessment of problem-solving in test conditions is also facilitated by staged processes. Quizzes and assignments that play a dual formative/summative role provide the exemplars necessary for criteria and standards-referenced assessment.

Some of the most valuable contributions of tests and examinations to student learning are the elements of culmination, synthesis and overview involved in the preparation by students. This can be greatly assisted by allowing and encouraging student-prepared notes to be used during tests and examinations. There is also one aspect of quizzes, tests and examinations that is essential for statistics, namely correctness and avoidance of vagueness, whether in question, output, or expected response. Statistics may be the science and art of data investigation, modelling and interpretation of uncertainty, but it is also a systematic, quantitative and coherent discipline with correct and correctly applied concepts, principles and procedures. Imprecise or vague questions that may be appropriate for informal discussion have potential for long-lasting damage to learning if used in a test situation.

Similarly, multiple-choice questions can play a valuable role in learning and assessment, provided they are balanced by other forms of assessment involving short response items, choice and interpretation of procedures, problem-solving and synthesis. Wild *et al.* (1997) describe just some of the ways in which they can be used effectively and efficiently. Choosing statements that are correct, incorrect, appropriate or inappropriate is of great value to student learning in statistics with its combination of concepts, systems and interpretations. Part marks can also be very effective in multiple-choice systems. Ambiguity should be strenuously avoided; one of the most educationally-unsound assessment techniques is asking students to choose the least ambiguous response, with no part marks.

2.10 Writing and critiquing writing about statistics

These are areas with potential for variety in types of assessment and types of statistical learning and thinking. Apart from the critiquing and writing of media reports (Watson, 1997; Gordon *et al.*, 2008), there are essays and other forms of presentations (for example, web pages) that can be researched and written, for example on the historical impact of statistics on science, medicine, industry or society in general. Another slant altogether is provided by students' reflections on their learning (Bulmer and Low, 2008). Awareness of the possibilities of assessment for learning in these areas is starting to grow.

2.11 Concluding remarks: Balance of variety

The importance of variety in assessment in statistics is threefold: variety in authentic contexts and data sets is needed to develop statistical thinking; variety in types of assessment is needed for the extent and balance of concepts, principles, procedures, interpretations and nuances of statistics; and variety is needed for the diversity of students and their learning styles. All types of assessment can be designed so as to contribute in at least some way to learning. However, too much variety in assessment, whether formative or summative, runs the risk of overloading the learning experience; individual components should complement each other, each having a role in an overall integrated, balanced, developmental and purposeful learning package. Assessment is aligned with learning outcomes in an iterative and ongoing process that asks of each assessment component, what is of value that is being assessed, and of each outcome, how it is learned and assessed.

The variety and extent of demands and pressures on university assessment can sometimes appear overwhelming and even contradictory. Whilst balancing formative, summative, flexible, continuous, rich and authentic assessment with demands for criteria and standards-referenced assessment, and developing generic graduate capabilities such as teamwork, problem-solving and communication skills, the problems of over-assessment and the politics of pass rates and attrition must be considered.

Selecting, developing and implementing appropriate types of assessment and variety of contexts and learning experiences for a learning and assessment package in statistics require a range and variety of knowledge: of the students and their backgrounds and programmes; of the statistical objectives and their place in the statistics spectrum; of how statistical thinking is learned; and of the work of statistical professionals, education practitioners and researchers. Knowledge of types of assessment and their roles in learning across the areas, dimensions and levels of statistics can come only from the sharing of experiences, innovations and their effects, reported in all possible forums, both verbal and written. Such discussion should be ongoing and far-ranging and bring together statisticians and statistical educators across interests, workplaces and nationalities.

3

Assessing for success: An evidence-based approach that promotes learning in diverse, non-specialist student groups

Rosemary Snelgar and Moira Maguire

3.1 Introduction

In this chapter, we discuss our experiences at the University of Westminster of developing an evidence-based approach to assessment that enables differentiation on a Masters programme in Psychological Research Methods. The assessment strategy is designed to promote effective statistical learning in non-specialist graduates studying this course. Statistics and quantitative data analysis are taught together with psychological methodologies. Although our students have some experience of quantitative analysis, they have very heterogeneous statistical skills, aptitude and experience, and some struggle in this area. Thus, the need for differentiation was an important driver in developing the assessment strategy, both for the course as a whole and for individual modules. Central to the assessment process is the explicit use of both formal and informal assessment and feedback. Despite the general focus on formal assessment, evidence suggests that informal assessment may be more important in developing statistical thinking. We built in structured opportunities for informal assessment and feedback without adding to workload. Formal and informal approaches, integrated from

Assessment Methods in Statistical Education Edited by Penelope Bidgood, Neville Hunt and Flavia Jolliffe

the beginning, are: very early assessment and feedback; a 'staged' approach to quantitative projects; and generalisation of feedback.

3.1.1 The need for differentiation

The expansion of higher education within the United Kingdom has posed a number of challenges to educators, not least the matching of teaching and learning strategies to large and diverse student groups. These issues can be particularly acute for those teaching statistics. Much statistics teaching is 'service' teaching, delivered to students on other programmes, such as social sciences or business, often as part of research training. These service teachers may not be statisticians themselves. This is certainly true of psychology; the statistical content is usually taught by psychologists rather than statisticians, a case of non-specialists teaching non-specialists (see Boynton, 2004). In terms of those studying statistics in higher education (both undergraduate and postgraduate), many are not doing so through choice; they have chosen to study another discipline. The statistical content of non-specialist courses can be daunting for these students (Boynton, 2004; Mulhern and Wylie, 2004). Lack of motivation and statistics anxiety are recognised as barriers to statistical learning and teaching in psychology students (Conners *et al.*, 1998) and in other disciplines (Onwuegbuzie and Wilson, 2003). Performance extremes are another challenge for educators (Conners *et al.*, 1998), making it difficult to pitch tuition effectively and fairly. Here we are specifically concerned with psychology postgraduates studying psychological research methods, but the issues raised are common to many other disciplines.

3.1.2 Statistics and psychology students

Quantitative research is central to psychology. Statistics and data analysis are usually taught together with research methods. This has many advantages, by anchoring the statistics and providing a conceptual framework. Student engagement is more likely, since they can clearly identify the usefulness of statistics in 'doing psychology'. Empirical research including statistics is identified as a core knowledge domain by the QAA (the Quality Assurance Agency for Higher Education in the UK) in its benchmark Statement for Psychology (QAA, 2007), and is also required in the British Psychological Society Core Curriculum to confer the Graduate Basis for Chartered Membership. The entrance requirements for psychology degrees at UK universities generally stipulate a relatively low minimum level of mathematics achievement and, although many students exceed this, research methods (or statistics) is the area in which our students are most likely to struggle. Indeed Mulhern and Wylie (2004) provide evidence that psychology undergraduates may be poorly prepared to study statistics. They compared two cohorts of psychology undergraduates (1992 and 2002) on six components of mathematical reasoning relevant to statistics (calculation, algebraic reasoning, graphical interpretation, ratio, probability and sampling, and estimation). The performance of the 2002 cohort was significantly poorer than that of 1992. They also reported generally poor knowledge of square root and manipulation of decimals. This illustrates the problems in the level of prior knowledge and skill that can be assumed when teaching statistics to non-specialist undergraduates. Mulhern and Wylie (2004: 356) go so far as to say 'the teaching of quantitative methods to large, heterogeneous groups of students is our greatest pedagogic challenge within Psychology'.

Although the issue should be less acute when dealing with postgraduate students, our experience is that they are diverse in terms of prior statistical knowledge, skill and motivation. This is not uncommon; Conners *et al.* (1998) suggest that it is a consequence of statistics not being adequately embedded into the undergraduate psychology curriculum. Furthermore, while all our students have a degree in Psychology or a related discipline, some have studied in other countries or some years ago. So, while we can make some assumptions, experience has shown that we need to review basic material on this course. That said, some of our students are statistically sophisticated and we also need to accommodate them.

3.1.3 The Masters programme

Our intention for the MSc in Psychological Research Methods is to provide an academic environment in which a wide range of psychological research methods is taught at postgraduate level, the needs of preparation for professional training and employment in Psychology are met, and several areas of application are considered. By 'research methods' we mean the whole research process: designing studies; conducting studies (data collection); analysing data; interpreting the data analysis; and reporting studies. We want students to gain knowledge and skills that will prepare them to make the most of future doctoral studies or other training. We also want them to develop their facility for self-directed learning, personal development and career management. The overall course aim is to provide training and to facilitate the development of skills that will enable the student to carry out and report original research of high quality in psychology. Given the nature of the course, a founding principle for student learning was that they would conduct their own research. Thus, students choose a topic, not just for their dissertation but also for each methods module. There are certain constraints: of ethics, practicality, and using methods appropriate to the particular module. Nonetheless, students are enabled to conduct studies in a wide range of psychological areas, and have commented very favourably on this aspect of the course. This use of research projects is expanded upon below.

Course structure The course comprises six taught modules and a dissertation. Four of the taught modules cover methods of data collection and data analysis; two for quantitative methods and two for qualitative methods. The qualitative modules do not include statistical data analysis, of course, but do require a suitably rigorous approach to the process of data collection and data analysis. General principles can thus be cross-applied between these four modules and also with the dissertation.

Quantitative methods modules Both quantitative methods modules deal with the whole research process, or the enquiry cycle, of research design, data collection, data analysis and interpretation. Data analysis topics start with a review of descriptive statistics, probability, and statistical inference, and then address the general linear model. Issues such as sampling, assumptions, approaches to modelling, and data screening and management are covered. We cover a range of multivariate and other data analysis techniques. The software packages used for statistical analysis are Statistical Package for the Social Sciences (SPSS), and Analysis of Moment Structures (AMOS) (Arbuckle, 2008) for structural equation modelling.

Use of projects The statistical and data analytic content is taught within the context of the key relationship between research design and analysis. Understanding is facilitated through the small research projects that students devise for their chosen topic. Recent examples include: 'the effects of word structure on recall'; 'anxiety responses to news reports on terrorism'; 'the relationship between personality traits and charitable donations'. For each project students are required to develop an appropriate psychological research question, design and conduct an appropriate study to address it, and collect, analyse and interpret the data. Dealing with design and analysis together provides a framework for students to think about variation and for developing what Wild and Pfannkuch (1999: 227) refer to as 'transnumeration'. Particular attention is paid to measurement and sampling.

3.2 Development of the assessment strategy with reference to statistics

Our personal experience and the literature reviewed above indicated that the statistics component of the course would need particular attention. Recognising the need for differentiation (see Sections 3.1.1 and 3.1.2) was an important driver of the learning, and teaching strategies. Rather than treat the heterogeneity as a problem we hoped to develop strategies that would use it to enhance the learning experience of all (see Roback, 2003).

We are training our students to be independent researchers, so their own research and its assessment are central to the learning process. This influenced the assessment strategy for the course, and enabled us to integrate much of the assessment of statistical content into assessment of the research process via students' own quantitative research projects and dissertations (outlined in Section 3.1.3). As will be discussed in Section 3.4.2, informal assessment and feedback opportunities are built into the modules. The completed projects are written up as research papers and formally assessed. This approach means that the statistical content is applied and assessed within the context of a real enquiry cycle devised by, and so directly meaningful to, the individual student. Statistical thinking synthesises statistical and context knowledge to 'produce implications, insights and conjectures' (Wild and Pfannkuch, 1999: 228) and it is this synthesis, the research process and product, that is assessed. As the course developed we became more interested in the possibilities of matching the assessment and differentiation strategies and explicitly using the assessment process to support differentiation by meeting the needs of individual students.

Formative assessment is recognised as essential to development of metacognitive skills (Nicol and Macfarlane-Dick, 2004). The effects of feedback on student learning are positive and can be large (Black and Wiliam, 1998; Gibbs and Simpson, 2004). As described in Section 3.1.3, much of our formal assessment involves the students' own research, either as individuals or in small groups. Thus, feedback on any single research assessment is relevant to all the others and feedback on any assessment can have general applicability to other types of assessment. It is crucial that students recognise both the specific and general relevance of feedback; thus, we developed strategies to emphasise the generalisability of feedback and encourage transfer. Simply providing feedback is not enough. As Sadler (1989) explains, in order to benefit from it students need to be able to (i) know what they are aiming for, (ii) get feedback on discrepancies between goal and actual performance, and (iii) do something to resolve these. An important element of our assessment strategy was the provision of feedback that adhered to good practice principles (e.g. Gibbs and Simpson, 2004; Nicol and Macfarlane-Dick, 2004).

Formal assessment, whether it also has a formative function or not, is focused on assessing the extent to which the work meets the learning outcomes defined in the formal curriculum. Yet the whole learning environment is much more than the formal curriculum and some highly valued outcomes are more amorphous and difficult to specify. As regards statistics and data analysis, an important goal of the course is the development of statistical thinking and high level statistical literacy. We felt that informal assessment was likely to be particularly helpful in offering students opportunities to assess their understanding of statistical concepts and data analysis techniques in a non-threatening way. Students need to recognise informal assessment and feedback as such, and it should adhere to the same principles of good practice as formal feedback. Our assessment strategy focused on combining formal and informal assessment in a structured way to promote assessment for learning rather than assessment of learning.

The issues reviewed above led to the pillars of our assessment strategy:

- Building into relevant modules structured and explicit opportunities for informal assessment and feedback that do not add to the student (and staff) workload (Gibbs and Simpson, 2004).
- Using informal assessment to focus on learning rather than performance (Boud, 2000) and making explicit the link between formal and informal assessment.
- Encouraging students to recognise informal assessment and feedback (Nicol and Macfarlane-Dick, 2004).

- Encouraging students to use the information contained in feedback to monitor and modify their performance (Sadler, 1989; Yorke, 2003).
- Developing metacognitive skills, particularly self-assessment skills (Yorke, 2003).

3.3 Evaluation and development of assessment strategy

The assessment strategy is intended to:

- Promote the development of high-level research and data analysis skills and statistical thinking.
- Promote use of feedback to improve learning and facilitate metacognition.
- Increase motivation and engagement with the statistical and data analysis aspects of the course.

The course has been running since 2003, with one significant restructuring in 2006 to facilitate part-time students. As with most programmes, assessment practices have been modified at various points in response to particular issues or in order to better meet student needs. The strategy discussed here has been developed organically over a number of years. At each stage developments have been evaluated within the context of module and course evaluations drawing on a range of sources: formal and informal student comments; external examiner comments and reports; the views of teaching staff. We did not test specific interventions but instead implemented an evidence-based approach, drawing on the literature. Our approach to assessment is focused on learning rather than performance, a requirement of formative assessment (Boud, 2000). In terms of statistics and data analysis our approach is focused on developing students' skills in statistical thinking, in research design and analysis, and metacognitive skills, rather than performance per se. A further aim was to promote motivation and engagement with the statistical content. Statistical thinking and metacognitive skills are difficult to measure (Seabrook, 2006) and without an experimental approach, it is difficult to link changes to specific aspects of the assessment strategy. Much more straightforward is asking students about their use of feedback and to do this we distributed a short questionnaire to students just completing the course and to recent students for whom we had contact details. Since the course is small, this amounted to only 14 students and we received only six usable replies.

3.3.1 Very early assessment and feedback

The major transition from undergraduate to postgraduate work means that students both need and want feedback quickly, but this can be difficult when they have not covered enough material to assess. Our early assessment combines informal and formal approaches and acts as a diagnostic for staff and students. It occurs in the first semester core module in quantitative methods and analysis. The first class includes a self-test exercise on a straightforward data set. Students are required to enter the data and use SPSS to run the appropriate analyses for some straightforward questions of difference and association between two variables. After the exercise is completed, a tutor-led discussion provides students with immediate feedback about how they have performed. The fact that this is feedback is emphasised. They are then asked to analyse their own performance, clearly identifying sources of problems as (i) familiarity with the software, (ii) knowledge, or (iii) understanding. Once issues have been identified they use the information as a basis for remedial action, and student and tutor agree 'catch-up' strategies. For the tutor, the exercise enables a rough assessment to be made of each individual student's level. It also provides information about the overall level and composition of the group enabling us to tailor our teaching more effectively at a very early stage, an important function of feedback (Yorke, 2003).

In the first class students are also given information about the first formal assignment, requiring them to complete a brief psychological report (weight 5%), submitted in the third week and returned in the fourth week. We are able to give them constructive formal written feedback on their academic writing and the extent to which they are meeting postgraduate expectations. Dialogue is an important part of the feedback process (Juwah *et al.*, 2004) and we encourage this through small group discussion of feedback and performance.

From the tutors' point of view the introduction of the early diagnostic assessments, from which we obtain the level and range of each group, has been very successful and we have been able to use this information to plan and modify teaching. Another advantage is that we are better able to advise and support individual students; at a very early stage we can engage them in dialogue about their performance and our expectations.

The students have been consistently positive about the early assessments and feedback, although most are initially surprised at doing 'work' in the first day and having to submit an assessment after only three weeks. The small weighting of the initial formal assessment means there is little anxiety around it and student comments in class suggest that, largely as the result of the fast feedback, these early diagnostics are very motivating, promoting engagement with the course and with their own learning. The informal and formal feedback is used as the basis for structured in-class dialogue (that can be continued on a one-to-one basis). This means that by the end of the first and fourth classes the students have actively reflected on their performance and discussed actions to address the issues they and we have identified. This should develop metacognition and help them to implement improved approaches to study and assessment.

The questionnaire responses supported these informal comments from students. For the first diagnostic class, four of the six respondents agreed or strongly agreed that it had helped them to identify areas for revision and three of these indicated that they did in fact revise. For the short report, all but one of the respondents used the feedback when working on their next assessment for the module. All of those who reported using the feedback from the mini-report gained higher grades on the full report, whereas the only student who reported not using the feedback had a lower grade on the full report. Given the low number of responses to the questionnaire, no conclusions can be drawn but in the future we hope to further investigate potential outcomes of giving early feedback as described above, and also students' perceptions of its value.

3.3.2 A 'staged' approach to quantitative projects

Staged assessment is increasingly recognised as a useful strategy in helping students to use feedback (Nicol and Macfarlane-Dick, 2004). Although it can be difficult to manage in terms of scheduling and compliance with university regulations, for projects and dissertations staged approaches are straightforward to implement as well as very beneficial. It is common for students to submit a research proposal that is formally assessed and the feedback then used in running the project. As discussed earlier, research projects are a core teaching and learning method on this course. We were keen to develop the staged model to include both formal and informal feedback using an incremental approach that offers students the opportunity to apply feedback from diverse sources at many stages in their work.

Informal feedback is particularly important for the smaller in-module projects as the timescale does not allow for more formal approaches. Self-evaluations, peer feedback and traditional tutor feedback are all important. Structured opportunities to discuss informally or present progress are built into modules. These enable students to monitor and discuss their progress and get feedback that is recognised as such. Students are encouraged to review their work in the light of the discussion while preparing for the next informal or formal stage.

Early in each module, these informal assessment opportunities tend to focus on issues such as measurement and sampling error and determining sample size. As their projects progress, students need to draw on their knowledge of data screening and assumption testing, including concepts such as multivariate normality that they are often encountering for the first time. Concepts such as effect size and model fitting are relevant in many cases and at the analysis stage. In the case of the smaller in-module projects, informal feedback is generally verbal and comes from both tutors and peers. At each stage, students are asked to explain the reasoning behind their decisions. This is particularly useful in assessing understanding of concepts and enables tutors to address misunderstandings as well as provide guidance. A common problem for some students is inappropriate choice of data analysis technique. This prompts discussion of test selection issues, allowing misconceptions and poor conceptual understandings to be identified and challenged. Students use these opportunities to discuss their ideas and to seek advice. The real-world data that they collect means that they are often forced to consider issues such as what to do when the assumptions for a test are not met and how to make sense of unexpected results. Our students design diverse projects that in turn demand a range of different approaches to design, analysis and interpretation. A major advantage of the group format is that general issues, such as sample size and assumption testing, are considered in a range of concrete contexts making it more likely that general principles can be identified and understood. It also allows students to confront a wider range of specific problems than would be possible through their own projects alone (for example, around transformation of data to meet assumptions). The process is similar for the dissertation although in this case there is additional individual verbal and written informal feedback via the supervision process.

Where possible, in addition to staged informal feedback, we use staged formal assessment and feedback. For example, in the second quantitative methods module students conduct a group project, which is formally assessed in two ways. First, the group gives a conference-style presentation for which the grade and both verbal and written feedback are given immediately afterwards. Subsequently, each student writes an individual research report in which they justify their methods of data collection and data analysis, and report, interpret and discuss the findings. They receive formal written feedback from their tutors. In terms of statistical content this is often focused on interpretation and critical analysis of findings.

For many students, conceptual issues and understandings have already been clarified via the informal assessment and feedback.

The highly structured approach provides useful formal and informal milestones for students to monitor their progress against goals, get feedback and consider how to apply this feedback. It is motivating for all students and most relish the opportunity to discuss their own work. We have found the informal peer assessment elements to be particularly useful in terms of creating a stimulating learning environment. Those with stronger statistical skills tend to encourage and support those who struggle or are less confident. Weaker students benefit from the regular reviews as it enables them to break the tasks down into more manageable components, identify problems at an early stage and apply feedback when it is useful. Building the informal assessments into the teaching schedule means this does not negatively impact on staff and student workload. However, where informal assessments are built in, their success depends on student attendance. When they do attend they work very well but even at postgraduate level attendance can be an issue, not least because many students have substantial commitments to paid work.

The survey respondents were very positive about the staged elements, with all but one agreeing or strongly agreeing that the feedback was both used and useful (the exception was a neutral response to both these questions). The low number of respondents means that we cannot make meaningful comparisons between these responses and overall module and dissertation grades.

3.3.3 Generalisation of feedback

We encourage students to generalise feedback between assessment types. This began as informal advice, which was increasingly emphasised, and has recently been incorporated in the course handbook. A specific example is for an assessment in which students critically evaluate the methods of data collection and of data analysis in a recent published research paper; feedback is applicable both specifically, to reviewing papers for reports, and generally, to developing research skills and statistical thinking. As discussed earlier, students must be able to use the feedback and staged approaches to assessment are an exciting way of approaching this. However, in order to develop research and statistical skills it is important to look beyond the module and/or specific assessment and encourage broader self-assessment. Most evidence indicates that generic skills do not transfer easily (Billing, 2007) and of course the distinction between generic and specific skills is not clear-cut (Perkins and Salomon, 1989) as all skills are applied in some specific context. Nonetheless it is clear that transfer must be promoted. Helping students to develop their metacognitive skills is an important component of this (Cox, 1997). This promotes cognitive adaptability and flexibility in applying skills. The early diagnostics encourage students to reflect on their work and self-assessment is part of this.

Of the questionnaire respondents, four agreed that they generalised feedback across the course while one disagreed and there was one neutral response. All the respondents agreed or strongly agreed that the various small-scale studies developed their ability to think statistically, helped them to understand the relationship between design and analysis, and developed their understanding of the particular analyses used. Given the low numbers we were not able to examine associations between these responses and overall grades. We feel, however, that weaker students make less use of the feedback provided. In future, we wish to focus on such students, encouraging them to make appropriate use of the feedback.

3.4 Discussion

Our experience of explicitly using assessment to promote statistical learning has been largely positive. It has enabled us to meet the diverse needs of non-specialist students engaged in high-level study. Although applied in the context of a postgraduate psychology course, the strategies could be adapted for large undergraduate groups of non-specialists and service teaching on other postgraduate courses.

3.4.1 Early diagnostics

From our perspective the early diagnostics are the most useful aspect, particularly in terms of setting the context for statistical learning. In general, they enable us to be more responsive to our students and their learning needs. With large classes, structured informal diagnostic exercises (as in our first class) will probably be more feasible than the small formal assessment, in terms of resources. However, students could be asked to mark each other's work and the dialogue could occur within pairs or groups of students. Making the purpose explicit is a useful strategy for increasing engagement and motivation, and for developing self-assessment skills.

3.4.2 Staged assessments

Modular structures can make staged assessment difficult to implement successfully as there is little time for development, but it is possible to advance from the traditional 'two stage proposal plus dissertation' model by incorporating informal assessment opportunities. The feedback does not have to be individual or tutor-generated but can be generated and delivered via workshops, practice tests, and peer discussions (see Gibbs and Simpson, 2004, for suggestions). These provide non-threatening opportunities for feedback, without major resource implications. Structured informal assessment can also help overcome some of the difficulties associated with staged assessment on single modules, as they can take place even within a short timescale.

3.4.3 Generalising feedback

Using feedback and using it generally is essential if students are to improve their performance. Stronger students do tend to apply feedback generally; weaker students rarely do so. Motivating weaker students to use feedback is thus the least successful part of our overall strategy and we are looking at ways to address this.

3.4.4 Future directions

Despite our efforts, many students still do not use feedback effectively and weaker students in particular may produce work with the same problems time and time again. To be useful, feedback must be used. One way to make this more likely is to present the feedback without the grade (Gibbs and Simpson, 2004). We are considering releasing feedback only, on at least some assessments, and requiring students to respond before the grade is released, so ensuring that all students make use of early feedback and of staged feedback, and that they consider how feedback is generalisable to assessment within and between modules. In other modules we are piloting the use of implementation intentions to encourage students to apply feedback; Gollwitzer (1999) showed that forming implementation intentions (what, where and when) is effective in helping people to put their intentions into practice.

3.5 Conclusions

In this chapter we have described and evaluated an assessment strategy which enables differentiation. Early feedback, staged feedback, and guidance in generalising feedback can all assist non-specialist graduates in their statistical learning. The early diagnostic assessments have enabled us to address problems early on and to ensure that the content is pitched appropriately. Students value early feedback, which seems to enhance both engagement and motivation. The staged approach enables students to demonstrate that they have considered and applied advice in feedback, encouraging the development of generic skills. While students often recognise diverse sources of feedback they do not always use it effectively. We are currently exploring possible improvements, including asking students to respond to feedback before grades are released. Although applied on a postgraduate course, the strategies outlined here could usefully be adapted for undergraduate programmes to help meet the challenge of teaching statistics effectively to non-specialists.

4

Assessing statistical thinking and data presentation skills through the use of a poster assignment with real-world data

Paula Griffiths and Zoë Sheppard

4.1 Introduction

In his paper delivered at the 2002 International Conference on Teaching Statistics, Holmes refers to a 1960s Nuffield sponsored primary school development project that used an old Chinese proverb as its slogan: 'I hear and I forget, I see and I remember, I do and I understand' (cited in Holmes, 2002: 2). Holmes (2002) suggests that the idea of doing as part of the learning process is important for students learning statistics because getting involved motivates learning. At the time of this project, statistics teaching and learning in universities was very much based on doing by 'substituting numbers in formulae' (Jolliffe, 2007: 1), an approach to learning and assessment that resulted in students becoming proficient at answering questions similar to those they had practised in their learning, but did not necessarily aid statistics comprehension and application. In preparing students very little for life outside of the classroom, this approach is clearly inappropriate for students studying another discipline who wish to apply the principles of statistics.

Assessment Methods in Statistical Education Edited by Penelope Bidgood, Neville Hunt and Flavia Jolliffe

Many statistics teachers highlight how assessment drives what students will learn on a statistics course. Wild *et al.* (1997, electronic page 2) have characterised this as the 'what you test is what you get' phenomenon. Garfield (1994) draws out two principles often overlooked by examiners designing statistics assessments: the content principle and the learning principle. In relation to statistics teaching, the content principle covers the statistical content to be covered in a course and the learning (or process) principle acknowledges that assessment should incorporate appropriate development of concepts, insights, and ways of thinking.

Teaching statistics to students whose main discipline is not numerical can pose additional challenges, since they often arrive with low motivation and confidence (Jolliffe, 2007). Evidence from many studies has shown that, for these students in particular, the use of real-world relevant data examples brings meaning that can motivate and help them enjoy learning about statistical concepts and principles (e.g. Yilmaz, 1996). In more recent times statistical assessment has evolved to incorporate the use of real data in assessments (Jolliffe, 2007). However, Gal and Garfield (1997a) state that statistics teachers more commonly assess the doing of statistics and place less emphasis on communicative skills and making sense of data, which they argue to be important curricular goals. Kelly *et al.* (1997) further identify the importance of needing to know why a particular statistical procedure has been chosen, what the procedure is showing, and any limitations associated with interpreting findings. These skills have been described in the literature as statistical thinking (e.g. Chance, 2002) and may not easily be taught, but students can be given practice that nurtures the skills a statistician might use.

Zevenbergen (2001) discusses the lack of innovative assessment methods in higher education. Posters are used at meetings and conferences as a method of disseminating information and findings, and their use has transferred well to the school classroom setting (Baumgartner 2004: electronic page 4); Baumgartner suggests that students can achieve satisfaction from the authentic practice of creating posters by gaining 'ownership and pride in their product'. This chapter describes the use of a group poster assignment using real-world data to assess statistical thinking and presentation skills. It is one of the assessment methods used in a first year introductory data analysis module for approximately 50 undergraduate Human Biology students in a one-semester, 10-credit module (100 study hours) at Loughborough University, UK. The module is taught through a combination of lectures, practical classes covering questions relevant to recent lectures, and computer workshops teaching students how to use Microsoft Excel and SPSS to answer questions using real-world data relevant to recent lectures. Thus, the practical classes and computer workshops provide opportunities to practise statistical thinking skills.

4.2 Background to the assignment

Based on feedback from students suggesting that they found the module difficult, uninteresting and irrelevant to them as Human Biologists, but that the skills taught were important for conducting their dissertations, from 2003 the curriculum was redesigned to incorporate statistical thinking using real-world data. It was necessary to think about how these changes could be assessed to show the students the value placed on these aspects through giving a summative grade. To reflect this need, one of the major changes to the module assessment was to incorporate a group poster assignment using a real-world Human Biology database.

The database used contains 14 variables from a random sample of 100 Indian women. The data were selected from the publicly available National Family Health Survey (NFHS) (data are available at www.measuredhs.com), a nationally representative survey of households in India collecting demographic and health data (International Institute for Population Sciences, 2009). These data were chosen for the assignment because they are real-world data containing biological variables (e.g. haemoglobin levels, body mass indices), and therefore of relevance to Human Biology students.

The assignment aims to test three key learning outcomes:

- An understanding of why Human Biologists often use statistics in their research.
- The ability to manipulate data into a useful format for analysis.
- The ability to define a hypothesis and describe data considerations that need to be made for testing a hypothesis.

The assignment also tests subject-specific skills, i.e. the ability to:

- calculate basic statistics, including averages, percentiles, measures of spread, and confidence intervals
- identify different types of distribution
- test a hypothesis
- use statistical tables.

In addition, the assignment tests key skills, i.e. the ability to:

- develop team work skills
- develop critical thinking skills
- advance computing skills using software packages
- produce graphs and tables from data
- present results appropriately.

4.3 Description of the assignment

Students are asked to work in groups of approximately four and are given the NFHS data file in both Excel and SPSS format. They are told to produce a poster for the Indian Health Minister, answering questions using the NFHS data. The Health Minister was chosen as the audience for the poster to indicate the importance of the findings because they would be presented to a senior government official who would be familiar with the health concepts, but who would need the statistical findings to be explained in non-technical terms. The use of the poster format for presentation means that students have to be selective about what information to include and how to present it. Because posters are more visual than traditional reports, this assignment tests a different type of presentation skill. Learning to display information visually is a useful skill to acquire for use in the workplace.

Because both content and style are important for this assignment, students are awarded marks for both elements. The content assessment accounts for 75% of the marks and includes the accuracy, appropriateness and relevance of material presented. Style accounts for 25% of the grade, which includes poster appearance, whether it is attention grabbing, effective use of graphs and tables, and clear text including titles and labels. This division of marks has evolved over several years. The original split used had been 15% style and 85% content, but 15% was found to be insufficient to motivate students to give enough consideration to style. The percentage of the mark allocation to style was therefore increased to 25%, which has since motivated the majority of students to give this component adequate attention. Since the allocation of marks has changed, two out of 25 groups in the last two cohorts (2007/8 and 2008/9) have failed to achieve a passing grade for style and the mean mark for style has been 60%.

By assessing both content and style, the presentation of statistics can be assessed alongside content. For more anxious students this can mean that the focus is not only on the calculations, traditionally more feared, but also on the presentation of information about Indian women's health. The calculations still have to be undertaken; although because marks are not all orientated towards these, students begin to recognise that the class is not just about the content of statistics, but also the doing of statistics in relation to Human Biology.

The assignment is split into six tasks:

1. Appropriately graph some key variables and explain why data are displayed in this way (15% of content mark).
2. Choose and calculate appropriate measures of average and spread for selected variables while telling the Indian Health Minister what these statistics show and explaining why these measures are chosen (15% of content mark).
3. Construct tables appropriately to help the Health Minister answer certain questions and interpret what they show (15% of content mark).
4. Construct, graph, and discuss the limitations of a composite index using several variables relating to nutritional status (20% of content mark).

5. Tabulate this index by a socio-economic measure, appropriately group the index, and interpret what the table shows (15% of content mark).
6. Construct a hypothesis and choose an appropriate statistical test to test the hypothesis and interpret the findings from the test (20% of content mark).

The questions are designed to guide students towards the types of information to include on the poster, without dictating the exact methods to be used. This empowers them to decide on appropriate techniques to display the data, but at the same time requires them to give a rationale for why they have chosen to display information in this way (data presentation). The task of constructing a composite index requires students to think about how they combine information from more than one variable into one meaningful summary measure. This requires students to think about the meaning of each of the individual measures and to use their biological knowledge to evaluate how and whether the variables can be appropriately combined to produce one single index. Students are then encouraged to think critically about the measure (statistical thinking). The assignment therefore tests statistical thinking in terms of justification for methods used, interpretation of findings and limitations (Kelly *et al.*, 1997).

The breakdown of marks for each question is known to the students from the outset, and the marks awarded for each question reflect the content covered in the question and the level of statistical thinking required. Results from 25 groups over the most recent two cohorts (2007/8 and 2008/9) suggest that even the weaker groups of students can gain a good mark for Question 1, with none achieving a failure grade of less than 40% (mean = 82%). Similarly, Question 2 had two groups failing (mean = 72%). These questions cover the more basic material and students have had longer to practise these principles.

Question 3 resulted in two groups failing, but the mean was lower at 65%. Although some groups answered this question well, several struggled with the important real-life task of organising data appropriately into a table to answer particular questions. Three groups failed Question 4 (mean = 69%). This question incorporates more difficult calculation principles and statistical thinking, but has tended to produce good answers, possibly because this is the question that the teaching team receive most enquiries about. In response to questions, students are directed to the relevant lecture notes and book chapters. It is likely that because they perceive this question to be harder, students engage in the directed reading and this extra study helps them to produce good solutions. Despite the overall good performance on this question, there are some groups who are unable to construct a valid index. These are generally the groups that receive failure grades. Additionally, some are able to construct an index but are then unable to choose an appropriate graph format to display the data. Thus, they successfully engage in the more difficult content task of constructing the index, but fail at the conceptually simpler task of choosing an appropriate graph to display the data.

Question 5 produced the lowest grades with eight failures (mean = 48%). Students often inappropriately calculate percentages, do not adequately define their index or its grouping for the reader, and do not give a clear interpretation of their table. Students perceive this question to be easy and very few groups ask the teaching team for advice. Question 6 had seven groups fail (mean = 62%). Students can mostly construct a hypothesis. However, several groups chose an inappropriate test, and where groups chose the correct test, they often made calculation errors. The concept of testing a hypothesis is relatively new to students at the time of the assignment.

4.4 Supporting students to complete the assignment

Most students require extra guidance as they have not completed similar assignments in the past. Guidance is given through a lecture session providing advice on poster presentation and how to undertake effective group work; the session also allows students to form groups and ask questions. Some examples of posters of varying quality are displayed for students to view briefly.

Students are offered a two-hour drop-in session approximately three weeks after the assignment is distributed and two weeks before submission, leaving time for groups to start working on the assignment and to formulate any queries. The drop-in session has two postgraduate teaching assistants available who can guide students towards resources that can help them to construct their own answers. To address the lack of confidence the first postgraduate teaching assistants expressed as to their ability to direct this session, the module leader now meets the teaching assistants ahead of the session to communicate the kinds of skills that each of the assignment questions is designed to test and to share information on resources that might be useful to students.

4.5 Poster presentations

There are usually approximately 50 students split into groups of four, so that 12–13 posters have to be assessed. Since the 12 posters are timetabled to be presented in a two-hour time slot, each group is required to make a short oral summary of their poster to one member of the teaching team of four, not to the entire class. The whole group is required to attend the presentation session. The group can nominate one member to make the presentation, thus utilising the skills of more confident presenters, but each group member must be prepared to answer questions. This enables the teaching team to gain some understanding of each member's contribution, complementing the peer assessment that students also make if contributions are perceived to be unequal. A core set of questions has been developed to ensure that, for the sake of fairness, all groups are asked similar questions and have similar opportunity to show additional knowledge or to reflect on their experience of completing the assessment. There is a training session for the postgraduate teaching assistants relating to the questions to be asked. As well as focusing specifically on the content of the assignment, this generic set of questions also helps groups reflect upon their positive and negative experiences of working as a team and on how they might do things differently in the future. For example, students are asked if all group members participated in the assignment, how well they felt they worked as a team, what they enjoyed and did not enjoy about the assignment, how they would do things differently, and the most valuable thing they learnt. The list of questions asked by the teaching team has evolved over time, after assessing several cohorts of students and learning the common mistakes made on the assignment. The teaching team edits the question list each year, incorporating any new lessons learned from a cohort.

4.6 Student reflections on the assignment

In addition to the usual feedback for the module, in 2007/8 students were asked to complete a brief questionnaire to provide anonymous feedback on the poster assignment. A student class member collected in the questionnaires to give to the module leader. Most questions had a five-item Likert scale response: not at all; not much; average; very much; and a lot. Out of 48 students registered for the module in 2007/8, 46 were present when the evaluation forms were distributed, and 40 completed questionnaires.

Thirty-five per cent reported that the assignment had enhanced their data presentation and analysis skills 'very much' or 'a lot'. Similarly, 33% thought the assignment enhanced their poster presentation skills 'very much' or 'a lot'. Forty per cent reported that their team skills had been enhanced 'very much' or 'a lot' by undertaking the assignment. A lower proportion of the class reported enjoying undertaking the assignment, with 20% enjoying it 'very much'. Despite this, 80% recommended that a poster be used in the future for assessment rather than a multiple-choice class test (the 2007/8 class had also undertaken this type of assessment in the module). This perhaps signals that students do not equate any type of assessment with enjoyment, although, relative to other assessment types, the poster project is favourable. Levels of enjoyment could therefore be a source of variation that could explain differences in performance on the assessment.

4.7 Staff reflections on the assignment

The module is currently coordinated by two staff members assisted by two postgraduate teaching assistants. It is usual for the postgraduate teaching assistants to have similar levels of knowledge, training and experience, and normally they assist with the module for two consecutive years. They are trained through the university's professional development programme and also receive module-specific training from the module leader.

Although student responses indicate that a reasonable number do not enjoy completing the assignment, staff generally enjoy guiding students through this assignment. It is rewarding to observe groups applying the information they have covered in lectures, practical classes and computer workshops, begin to grasp the content principles as well as the doing of statistics. Staff members have observed that many student groups rise to the challenge of compiling a different type of assignment and take the opportunity to learn skills that they do not get tested on elsewhere. This bears out to the 'ownership and pride' displayed by students undertaking poster assignments, as described by (Baumgartner 2004: electronic page 4). Staff also find it rewarding to see the quality and amount of effort that go into some posters.

The assessment and marking of the poster assignment is more time-consuming than a traditional class test and probably equates with the marking of individual short laboratory reports. Students present their posters to one of the four-member teaching team over a two-hour period, meaning that eight person-hours are spent observing student presentations (12 groups). However, this session is extremely valuable to staff and students because it allows reflection on the learning process and the understanding it yields can be used to inform the teaching process for the next cohort of students.

In addition to the student presentations, the teaching team have a marking session which lasts 6–8 hours (24–32 person-hours in total). It is important that all members of the teaching team contribute insights gained from the presentation and questioning about the posters. Assessment of the poster style is subjective, so pooling several opinions enhances fairness in grade allocation.

The rigid marking scheme we used in the earlier years of the poster assignment was found to be extremely time-consuming to implement and we would find ourselves engaged in debate over whether a group deserved an extra half a mark. This marking scheme was also not as easily accommodating of additional information learned from groups in the assessment session, and the weighting of the different components, decided on the basis of the perceived difficulty of each task, could be open to debate. Therefore in 2008/9, a more flexible marking scheme was introduced whereby percentage marks that map onto degree classifications were awarded for each question. The main elements of the question that we had previously identified to be important were still considered but, rather than attributing a certain proportion of the grade to each of these aspects, one grade was allocated for the overall performance on the question – much as in marking an essay. This enabled any innovative responses given in the group assessment session to be incorporated into the percentage grade for the question, resulting in less debate over small mark differences. Although quicker to use, this system was more subjective.

Informal feedback on the content and presentation of the posters provided to the students during the assessment question session is followed up with formal, written feedback returned to the students when they are given their grades – and in time to inform learning for the final examination. Compiling this feedback adds approximately two hours to the assessment process.

Postgraduate teaching assistants have commented that they enjoy participating in the group marking session, as they gain experience of assessment and learn how and why marks are allocated as they are. At Loughborough University, PhD students can contribute to marking first year undergraduate work if they have been trained, but there are few opportunities for them to do this alongside more experienced members of staff. Although the person-hours used for marking are quite substantial and would probably not be possible to dedicate with a large class, having a single poster per group reduces marking time, because 12 posters are marked instead of 48.

It is inevitable with group work that some groups will experience problems. The module leader needs to be willing to work with them on resolving issues and will occasionally also need to act to ensure that reported and substantiated unequal contributions are reflected in the marks.

4.8 Discussion

This assignment has allowed both content and learning principles to be assessed by analysing and presenting real-world data, a practice that many others have shown to be important in statistics assessment (e.g. Jolliffe, 2007). By engaging in summative assessment of content as well as learning principles, students are more likely to attach value to these areas (Wild *et al.*, 1997). The assignment has been shown to fit with the curricular goals of the module and is manageable within the resources available. The clear marking structure and the feedback given to students is perceived to be fair. The assignment therefore fits the important characteristics for an assignment recommended by Wild *et al.* (1997). The assignment would probably be less appropriate for a larger class size or if postgraduate teaching assistants were not available.

This assignment also promotes statistical thinking, which incorporates communicative skills and making sense of data, identifying why a statistical procedure should be used, interpreting output from statistical procedures and expressing limitations of analyses – skills identified by others to be very important to assess in basic statistics classes (e.g. Kelly *et al.*, 1997). Chance (2002) argues that these skills are best acquired by practice. This assignment offers students the opportunity to practise their learning principles. By completing the assignment, students start to develop skills in designing effective posters. The assignment enables them to gain significant team-working skills in their first year of undergraduate studies. Student feedback reveals that a large majority value this experience and understand its importance.

Marks for the 2007/8 cohort of students completing the module show that a mean mark of 59% was attained on the first traditional multiple-choice class test assignment. In contrast the mean mark for the poster assignment was 67%, and this higher mean was maintained in the examination (65%). The higher mark on the poster assignment perhaps reflects that deeper understanding is gained by applying statistics principles through being hands-on with data. The fact that the mean mark in the examination remains higher suggests that the content information learned in the poster assignment is retained for the duration of the module.

The skills taught in this module are very relevant to the final-year project module, where students analyse data to complete dissertations. Staff who have been in the teaching group for a number of years have remarked that students who have engaged in the revised statistics module with the poster assessment are more confident users of data in their final-year projects than previous cohorts. Unfortunately, only this anecdotal evidence exists to support this idea.

We would recommend this type of assignment to other lecturers teaching statistics to non-statisticians. There are many discipline-specific, publicly available data sets that could be used to construct a poster assignment which would allow students to engage with data and statistical thinking relevant to their studies. The use of posters does not have to be confined to those teaching statistics; testing the generic skills to display information concisely and in a visually attractive output makes them relevant as an assessment tool for other disciplines. We are situated in a multi-disciplinary department and other lecturers in Ergonomics, Psychology and Human Biology are also now successfully incorporating poster-based assignments into their modules.

We have not succeeded in producing an assignment that everyone enjoys, but it does seem to be preferred to more traditional assignments. Students suggest that one way to improve enjoyment would be to work with a variety of data sets – not all are inspired by the Indian women's health data example. This would require preparation of more data sets and questions specific to each data set; students could then choose which data to work with. At the current time we do not have the resources to do this, although in an ideal world this could produce a more enjoyable assignment for the students.

Another weakness of the current assignment is that it takes more assessment time than multiple-choice tests and requires more training time for the postgraduate teaching assistants. Staff perceive these time factors to be justified by the clear benefits for all. The postgraduate teaching assistants gain from learning about the assessment marking process, and the training sessions equip them to rise to the challenge of answering students' questions about the assignment; hence their own self-confidence in using statistical principles grows. The current teaching team is considering introducing peer assessment of the style aspect of the assignment: student groups would have to critically evaluate what others had done, compare it to their own work and consider what attributes of style result in a top quality poster. This process potentially enhances the learning experience.

In conclusion, students have been enabled to reflect and learn from engaging with the poster assignment. As well as beginning to realise the value of data analysis and statistical thinking, they gain key skills such as teamwork and presentation. The poster assignment is therefore a valuable assessment tool.

5

A computer-based approach to statistics teaching and assessment in psychology

Mike Van Duuren and Alistair Harvey

5.1 Psychology and the statistics challenge

Psychology is one of the most popular subjects in higher education in the United Kingdom. It is the largest scientific discipline and the second largest academic discipline overall (QAA, 2007). It comes as a surprise to some new students that the subject has a substantial statistical component at its core and often requires a mathematics qualification as a prerequisite for undergraduate study. For BSc level entry in the UK, for example, students are required to achieve at least grade C at the national General Certificate of Secondary Education (GCSE) mathematics examination, taken at age 16. Despite this criterion a considerable number of UK psychology students experience a range of conceptual difficulties in mathematical thinking, which some authors claim causes significant learning difficulties in statistics modules (e.g. Mulhern and Wylie, 2004, 2006).

At the University of Winchester we piloted an ad hoc general numeracy assessment in 2006 and 2007, of all students in their first statistics class, to identify those who might benefit from extra-curricular numeracy coaching. The test covered the four core numeracy topics: arithmetic; fractions, decimals and percentages; descriptive statistics; and algebra. The resulting scores for the two undergraduate cohorts were entered into a multiple regression analysis, along with the same students' examination results for research methods and statistics

Assessment Methods in Statistical Education Edited by Penelope Bidgood, Neville Hunt and Flavia Jolliffe

modules, completed at the end of Years 1 and 2 (Harvey, 2010). The main aim of this study was to assess the overall predictive power of the numeracy test on measures of undergraduate statistics performance, with the further aim of determining which numeracy sub-topics were the strongest statistics examination score predictors. Unfortunately the numeracy assessment proved to be an unsuccessful measure. The arithmetic sub-component predicted a significant amount of variance in the Year 1 statistics examination scores, but only for the 2006 cohort. Beyond these data, our supposedly 'diagnostic' test was found to have no significant predictive power.

These surprising results contradict Mulhern and Wylie's (2004, 2006) claim of a strong relationship between general numeracy skills and statistics learning in psychology. However, there is some support for our findings in the psychology teaching literature. For example, Gnaldi (2006) found no relationship between mathematics qualifications on entering a degree, on the one hand, and undergraduate examination performance on an introductory statistics course, on the other; and Huws *et al.* (2006) actually found degree entry grades in the subjects of general science and English to be better predictors of overall psychology degree performance than mathematics entry grades.

Perhaps this finding of a weak relationship between secondary and tertiary level numeracy skills and undergraduate statistics ability is less surprising if one considers the contrasting skill sets required for the two subjects. General numeracy problems, of the form posed in our assessment, for example, tend to involve number manipulations and calculations that are somewhat abstracted from any wider, more meaningful context. They also entail the sort of 'number-crunching' that is today more typically performed by computers. Hoyles *et al.* (2002) refer to a contrasting set of skills, which they refer to as 'mathematical literacy'. These include:

- Systematic and precise data-entry techniques and monitoring.
- Context-dependent, multi-step calculations and estimations with the use of IT systems.
- The modelling of variables and their interactions.
- The interpretation and transformation of graphical data.
- The extrapolation of results across different domains; the identification of anomalous or erroneous results.
- The clear communication of these judgements and outcomes.

Thus, mathematical literacy describes the application of mathematics to real data outputs, work situations and practices (Hoyles *et al.*, 2002). We contend that these applied conceptual skills are extremely important for learning statistics and quantitative reasoning yet they are not well tested by general numeracy tests. So at Winchester now, rather than attempting to diagnose weak numeracy skills, we reassure students that any prior mathematical difficulties they may have encountered will not necessarily disadvantage them in statistics. We also stress that a more important indicator of course success is the ability to integrate newly acquired statistical concepts and procedures into their wider psychological contexts and the students' own general knowledge. This is facilitated by the design and integration of our research methods modules throughout the undergraduate psychology programme.

5.2 An integrated research methods syllabus

As with all undergraduate psychology degree courses accredited by the British Psychological Society (BPS), research methods and statistics modules are compulsory for the first two years of the Winchester programme. However, we offer an optional advanced statistics module in the third and final year of study. The changes in learning outcomes across the three years of study are commensurate with the expectation that students develop their conceptual understanding of statistics and progressively integrate it with more advanced analytic procedures. There is also an expectation that students gradually extend their knowledge of procedures and concepts to wider contexts including, for example, peer-reviewed journal articles, statistical reports in more popular media and, not least, their own empirical endeavours.

Students on all statistics modules receive a series of lectures immediately followed by computer-based practical workshops, where they have the opportunity to demonstrate their understanding of the statistical principles and methods taught using the software package SPSS. In Year 1 of the statistics curriculum, students are introduced to the ways in which the observation and recording of human behaviour lead to numerical analysis and interpretation. With the use of small data sets (usually no longer than half a side of an A4 sheet) the emphasis is on summarising and understanding different data types descriptively, at both a graphical and statistical level. In addition a number of simple inferential tests are introduced, with single-factor ANOVA being the most complex.

In Year 2, this foundation of knowledge is extended with simple regression and inferential testing applied to multi-factorial designs. Importantly, students in Year 2 are introduced to the wider context of data analysis, which includes not only more complicated relationships between ethics and design but also the ability to critically evaluate the relationship between a given design and the questions it purports to answer.

The final year (optional) statistics module covers more advanced multivariate techniques including: multiple regression; factor analysis; and structural equation modelling. Third-year students are expected to show a more sophisticated evaluation of a given design and data set, taking into account more design features and basing their evaluation on wider knowledge in a more sophisticated way. They would also be expected to show a more advanced understanding of key conceptual issues in statistics, such as the difference between significance level and effect size.

Despite some uniformity in the delivery of statistics material across the three years, the curriculum is characterised by a pedagogical philosophy in which students are shepherded through a gradual transition from dependent to independent learners. This approach involves much 'hands-on' teaching during Year 1 statistics practical sessions, followed by less direct tutorial instruction and an increasing emphasis on independent practice as students' progress through Year 2. By Year 3, students should rely minimally on formal practical instruction and are expected to demonstrate a greater degree of self-motivation, initiative and problem exploration in their learning.

The main emphasis of the statistics modules described above is on theory and the implementation of methods using SPSS. However, these sessions are taught in conjunction with research methods modules where students learn how their newly acquired statistical skills are implemented in the practice of conducting their own empirical research. Topics covered in these modules include experimental design, sampling, data collection, analysis and report writing. In Year 2 research methods classes, students are further required to critically evaluate a number of published empirical findings drawn from a range of psychology journals. The culmination of statistics and research methods training occurs in a final year module in which students apply their skills to their own independent empirical project.

5.3 Approach to statistics teaching

As mentioned previously, our statistics sessions are based on an integrated mix of lectures and follow-up workshops. While lectures are largely expository, students are quizzed on the material throughout. They are also sometimes encouraged, either individually or in small groups, to make specific connections between new and previous content, such as how different test assumptions relate to each other. During workshops students engage in a variety of self-paced activities in which they can freely revisit the lecture slides on their computer screens, or consult the module study-guide, in which the content of the lectures are written in a more 'chatty', informal style and embedded within additional examples. The study-guide also includes weekly worksheets comprising practice exercises. These worksheets provide students with an opportunity to apply the statistical knowledge and procedures covered in lectures to small-scale psychological studies. Students may also be asked to create data to demonstrate their understanding of a concept. For example, one task may be to devise a data set that approximately results in a given correlation coefficient. As an incentive to complete these exercises, students are awarded 0.5% towards their final examination grade for each (of a maximum of 10) completed worksheet. These assignments are self-marked with students comparing their work to model answers supplied the following week. Lecturers further facilitate the workshops by circulating the class offering help and encouragement to students on an individual or small group basis.

Our teaching of statistical concepts is further enhanced, wherever possible, by the use of analogy. This approach is of course not new, but recent empirical evidence suggests that offering analogies from more familiar or applied knowledge domains may be particularly effective when teaching statistics to less mathematically inclined students. Zeedyk (2006) found that mapping various statistical concepts and procedures on to a police detective work analogy enabled students to assimilate and remember new statistical concepts more effectively, both in terms of assignment outcomes and qualitative student evaluations (see also Martin, 2003). To this end, we have exploited the Internet as a means of communicating analogies for statistical concepts. For example, in order to explain an interaction effect we have employed the analogy of a car differential. This was illustrated using an animation taken from a car DIY web site (http://auto.howstuffworks.com/differential.htm) showing the effect of the engine's force (i.e. the experimental effect) being distributed differentially across the four wheels (i.e. representing the levels of different variables) when turning a corner.

Other examples might explore the principle of the increase or decrease of an F-ratio to an increase or decrease of a signal-to-noise ratio, where the audibility of the tutor's voice (signal/experimental effect) varies greatly depending on where the tutor is located in relation to the static background noise and student seating position in class (noise/statistical error), which can be demonstrated with a web-based film clip at www.classroomhearing.org/acoustics.html.

For statistics lectures in Years 1 and 2, all the methods taught are contextualised and students are given the opportunity to apply them immediately afterwards. However, Year 2 students, in accordance with an increased emphasis on the critical evaluation of methods at this level, are given additional bi-weekly research methods classes in which they learn to evaluate a range of published empirical studies. These are carefully selected from the psychology literature on the basis of two criteria. Firstly the methods employed in each paper must dovetail with the content of that week's statistics lecture. Secondly the articles selected for evaluation, though published, must be relatively weak in terms of one or more of the following features: rationale for study; logic of argument and conclusions; experimental design/method, or statistical analysis; and reporting of results. These critical evaluation sessions require students to read and then in small groups prepare an evaluation of one of the selected articles, after which the tutor facilitates a discussion around the evaluation of the study. These tutor-led discussions should take in a wide range of issues, such as whether the purpose of the study was clear and well-justified, whether the research questions were appropriately explicated and theoretically driven, and whether the choice and quality of the design and statistical methods employed were appropriate given the nature of the data. The assessment for this part of the module comprises of an essay in which students must critically evaluate a published study selected by the tutor. These exercises allow students to judge statistics in a real-world context whilst building their confidence in evaluating published material, to which they may otherwise have been overly deferential.

For Years 1 and 2, further attempts are made to contextualise the statistics problems students are given by including, where possible, versions of research designs that have been applied by the tutors in their own areas of expertise. The data from these examples may also have been alluded to in other modules, for example, developmental or cognitive psychology. To save time (in both class and examination contexts), only small data sets accompany these designs, which students key into SPSS themselves. However, in the optional advanced statistics module in Year 3, we invite students to use the Internet to find far larger (social science) data sets. They are then encouraged to generate testable hypotheses and accompanying analyses from these data. One example is the European Social Survey, which we have used extensively (www.europeansocialsurvey.org) for Year 3 statistics teaching.

5.4 Statistics teaching resources

Typically, statistics classes are not an aspect of the psychology degree programme that most students enthusiastically anticipate and a prior struggle with secondary school mathematics may be at the root of this negative perception (Mulhern and Wylie, 2006). It is for this reason that we remind students, as suggested in the introduction, that failure in one of these subjects does not necessarily mean failure in the other, as many of our students have demonstrated. In a further effort to counter negative anticipation and to assist learning, we recommend to students a number of additional statistics resources, which we believe are well written, but in a light-hearted and engaging style. A good example of such a recommendation is Field (2009a, third edition), which won the prestigious BPS Book Award in 2008 – a unique achievement for a statistics textbook, as far as we know. Field's text is extremely popular with students and its self-deprecating style and frequent use of humour are often cited as reasons for its appeal. The use of humour may be effective because it relieves mathematics-anxiety, while simultaneously improving learning through increasing the amount of attention students pay to the material, which in turn should lead to improved memory retention and recall (Field, 2009b). This may be true but, besides the use of humour, Field has also gone to great lengths to find rather bizarre examples and data from the psychological literature from studies that nevertheless address important theoretical questions. One such example is the data from a study in which lap dancers were recruited in order to examine a question about the evolutionary benefits of attractiveness (Miller *et al.*, 2007).

In addition to a range of textbooks and handouts, our second-year students are also advised to look at some of the many excellent (and free) interactive web sites and commercially available software packages, such as *Statistics for the Terrified* (2008), or Richard Stephens and Sol Nte's (2008) ingenious interactive visual workspaces to aid the understanding of analysis of variance (see http://www.psychology.heacademy.ac.uk/miniprojects/anova/) and normal distributions (http://www.keele.ac.uk/depts/ps/RSStat/index.html). The ANOVA exercise displays two dynamic overlapping normal distribution curves, descriptive statistics, an ANOVA summary table and links to a number of tutorials. Among a host of other features, the curves can be manipulated so users can see directly how the changing shape of the distribution impacts on the results of the ANOVA.

Resources such as these offer a playful means by which students can develop an intuitive appreciation for how numbers in different statistical contexts relate to each other, rather than passively rote-remembering seemingly arbitrary procedural steps, as was so often the hallmark of the statistics learning experience in the past. Many of these web-based interactive statistics packages also include games and quizzes that help sustain students' attention throughout their learning. As a further means of providing extra-curricular statistics support, we pay a small number of our best third-year statistics students to facilitate weekly drop-in surgeries designed to support students in Years 1 and 2 with specific statistics-related problems. This also provides an opportunity for the Year 3 students taking part in the peer tutoring scheme to extend their own statistical knowledge through teaching.

5.5 Statistical literacy

In addition to gaining confidence in understanding and applying the statistical methods learnt, students also need to develop their 'statistical literacy' (e.g. McIntosh *et al.*, 1992; Watson and Callingham, 2003; Frankcom, 2007). This refers to the ability to apply statistical, mathematical, and linguistic skills to 'the various agendas which may be absent in statistics classrooms or in empirical enquiry contexts' (Gal, 2002: 15). We aim to foster statistical literacy at Winchester by providing students with opportunities to critique both published scholarly articles, as described in Section 5.3, and statistical procedures and results reported in the wider media. The development of this latter skill is particularly important in a world where public opinion is still far too easily swayed by unscientific and implausible arguments, assertions and evidence.

This is very much a concern for Ben Goldacre, who has been naming and shaming purveyors of dubious statistical practice in the *Guardian* newspaper since 2003. These 'Bad Science' columns are also archived, with reader comments, on the author's own web site (www.badscience.net). Goldacre, a medical doctor and scientist, won the Statistical Excellence in Journalism Award of the Royal Statistical Society for one of his columns entitled 'When the facts get in the way of a story' (Goldacre, 2006) and explores, in a highly engaging and jargon free manner, the background to myriad 'scientific' claims reported in the media. The critical commentary given in that article, and in his book (Goldacre, 2008), to a wide range of popular topics – including complementary and alternative medicine, product marketing, bogus dyslexia cures, television nutritionists, anti-immunisation campaigners and 'brain-training' programmes – make this a highly relevant and entertaining resource for statistics students at all levels.

Over the course of a semester at Winchester we have also, on occasion, organised little competitions where the aim for the students is to collect their own examples of bad statistics reporting (from newspaper cuttings, TV and radio quotes, blogs, etc.) and explain their rationale for selecting the pieces (i.e. their alleged basis of inadequacy). These cases might involve examples of sampling problems, confusion between inferential and descriptive statistics, distortions in graphical representations of data, the interpretation of correlation as causation, faulty logic, selective reporting of results, or simply a failure to provide sufficient data to justify the claims made. Such examples of poor practice are often heard in radio interviews, which are now frequently available as streamed or downloadable audio files stored on radio station web sites.

5.6 Statistics examination, revision and assessment

Our general approach to the assessment of statistics performance is to employ both formative and summative methods. A formative assessment is given in the critical evaluation component of the Year 2 statistics module, whereby extensive feedback is given to all students on the quality of their essays. The examinations, on the other hand, are summative modes of assessment, although students who fail to pass first time are invited to discuss their performance with a tutor before being given their one opportunity to re-sit (but only for a maximum grade of 40%).

For the final teaching session in the first- and second-year statistics modules students are given an examination revision class where an overview of the examination format is provided. More importantly the revision class is used to demonstrate a systematic strategy that will enable students to 'diagnose' each research problem posed in the examination, and which can be applied independently of how a design or accompanying data table may look. We go through a past examination paper to remind students how to identify: the research design; the independent and dependent variables; the levels of the independent variables; the type(s) of data shown; and the most appropriate test(s) of inference to address the given question. They may be asked to consider whether the design posed in the question is within – or between – subjects, or whether a test of association or difference is required. We find that this approach helps mitigate against a tendency by some to map (sometimes superficial) design features from previous examples (such as the way the data were tabulated) onto new designs as a means of determining which test to apply. Thus, the goal of our revision sessions is not to complete the actual data analysis required for the questions but to state, for each, the correct diagnosis and its accompanying rationale.

Collaborative learning is fostered in the revision sessions, with students encouraged to work in pairs to pose diagnostic questions for each design description. In fact, our degree programme is guided by a pedagogical ethos that acknowledges the importance of facilitating collaborative learning, wherever possible, across the entire programme. For a number of years we have promoted the use of Student Support Units (SSUs), a voluntary scheme aimed to promote the idea that learning can be facilitated when students organise themselves into small groups in order to benefit from contrasting learning styles and experiences. Under the guidance of a tutor, SSU membership can offer students both emotional support and pedagogical assistance on topics as diverse as academic reading and writing skills, presentation practice, reciprocal peer tutoring and, particularly statistics examination revision (Barton *et al.*, 2006, 2007).

The examination revision tutors also suggest a number of exercises that are conducive to work in these small groups in and out of the class context. In one example, the traditional 'here is the context, here is the data, test the hypothesis' (Hubbard, 2003: 1) scenarios are reversed. Students gather in small groups in which each member formulates and writes down on a card the framework of a design for an empirical study. This might include only the number of independent variables, their corresponding number of levels, and data types. The cards are then placed in a box and one is drawn at random. The group's task is to then 'flesh-out' this skeletal design by thinking of appropriate labels for each variable and its levels. Group members must then think of specific hypotheses to formulate around this design and create a small (hypothetical) data set to accompany it, which is then analysed using SPSS. Previous work elsewhere has shown that students asked to produce hypothetical data in this way gain a deeper understanding of both the procedural and computational features of the analysis (e.g. Hubbard, 2003).

The statistics examinations in the first two years of study follow a broadly similar format. They consist of four questions of equal weighting with each offering a short description of an experimental design with an accompanying data set. These data sets are small and candidates are required to enter the data into SPSS themselves. In the Year 1 examination, candidates are required overall to: identify data types and variables (both independent and dependent); evaluate experimental designs and hypotheses; provide and interpret appropriate descriptive statistics, tables and/or graphs; identify, conduct and report appropriate analyses to test given hypotheses (referring to test assumptions); and draw sensible conclusions from the results. For Year 2 more complex experimental designs are introduced into the examination (e.g. mixed designs, factorial ANOVA, planned and post-hoc comparisons, tests of association/correlation, and data transformations) and candidates are expected to demonstrate greater levels of critical analysis. In Year 3 students are required to analyse fewer data sets but the extent of the analytic task is greater. For example, candidates may be expected to offer a critical evaluation of the design that addresses deeper theoretical concerns or wider contextual considerations. Year 3 examination questions are also associated with much larger SPSS data files, which students access via their examination accounts.

All statistics examinations at Winchester are three hours in duration and are what we term 'open-book', meaning that candidates are permitted to take into the examination hall any textbooks and lecture notes they wish. In the revision sessions we emphasise that it is better to take just a few sheets into the examination hall, including a list of diagnostic questions to help interrogate the research scenarios in each question. Many candidates nevertheless bring additional materials, including annotated text books and lecture worksheets. Although, realistically, there is relatively little time for candidates to consult this supporting material during the examination, simply permitting it into the examination venue seems to alleviate examination anxiety.

As mentioned previously, the statistics examination is administered via a desktop computer to which each candidate is assigned an examination account allowing access only to Microsoft Word and SPSS. Email accounts, the university network, the Internet and other software are strictly barred. Although technical problems occasionally arise in this form of computer-based examination (e.g. SPSS crashes) they are usually minor, are resolved swiftly and affect only a small minority of candidates. The time lost fixing the fault is always recorded by an invigilator and the affected individual is then invited to take this as extra-time after the official finish. After the assessment the contents of the examination accounts (i.e. Word files, SPSS data and output files for each candidate) are collated by IT services and sent to the tutors on CD ROMs. Tutors then grade the examinations, usually in the same room, to maintain consistency in marking calibration.

The computer-based approach described here is certainly a more time-consuming method of assessing statistical understanding than traditional examinations, where SPSS outputs for each question are provided to students in paper form. However, it is a more penetrating approach, in which both theoretical and procedural knowledge is examined. The computer-based assessment also mirrors the situations in which students are likely to employ their analytic skills in subsequent postgraduate and professional positions. In these contexts, efficient data management, accuracy and speed of analyses are generally viewed as integral features of the data analytic task. One could even argue that managing (unanticipated) technical problems under limited time constraints is a further skill students may one day value in the real world of statistical problem solving.

5.7 Concluding remarks

For statistical literacy to be developed effectively, we conclude that students should be given a variety of learning experiences across a triad of domains within the teaching curriculum (see Figure 5.1). The first of these is variety in the student's experience of the context in which statistical information is encountered. The second consists of variety in the resources provided for students to learn from and interact with. And the third consists of variety in assessment opportunities for students to express and demonstrate their growing statistical literacy.

In terms of the specifics of this chapter and the approach taken at Winchester, variety in encountering statistical contexts is provided by published psychology journal articles, tutors, reports in the wider media, and even the students themselves. The varieties of resources through which students may interact and learn include contrasting oral and written delivery styles, selected web-based statistics tutorials and quizzes, electronic archives, and more experienced student demonstrators. In terms of differing forms of assessment, our examples include opportunities for peer-assessed weekly worksheets, quantitative data analysis,

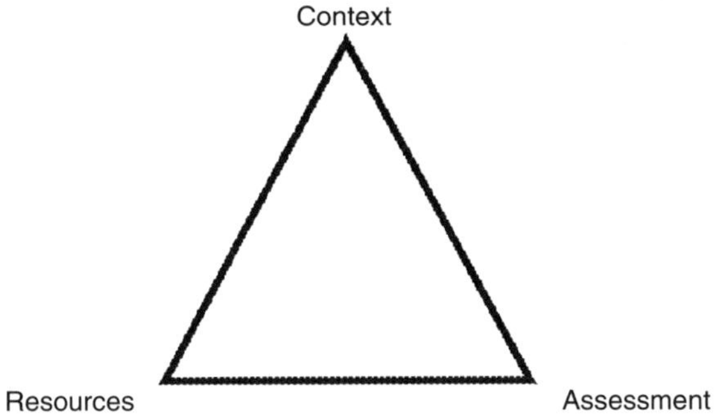

Figure 5.1 A triad of experience-domains in statistics learning

qualitative appraisals of published empirical designs and statistics with each year's statistics teaching culminating in a real-time, computer-based examination.

Despite these efforts, it remains a challenge to engage students with statistical concepts in non-trivial ways, especially those whose interest in psychology was not initially motivated by a passion for experimental science. It is our contention, nevertheless, that this challenge is best met by integrating statistics learning across the entire psychology syllabus and wider media, and by revealing to students the full scope, power and importance of statistical methods in society. Students who grasp this, and who develop an interest in the use (and abuse) of numbers around them should then, we hope, be inspired to learn the more prosaic statistical skills required to become competent social scientists.

Part B

ASSESSING STATISTICAL LITERACY

6

Assessing statistical thinking

Flavia Jolliffe

6.1 Introduction

The terms 'statistical thinking', 'statistical understanding', 'statistical reasoning' and 'statistical literacy' are used a great deal these days, to some extent interchangeably. We might want to assess any or all of them and also, as traditional types of assessment have attempted to do, statistical knowledge. It might be argued that knowledge underpins thinking, understanding, reasoning and literacy. On the other hand, knowledge cannot occur without understanding and thinking.

How then are these terms defined? In Chapter 7, Garfield *et al.* give definitions of statistical literacy, reasoning, and thinking and place the three in a kind of hierarchy. They claim that statistical thinking involves a higher level of thinking than statistical reasoning and that statistical reasoning involves understanding concepts at a deeper level than statistical literacy. According to Garfield *et al.*, citing Rumsey (2002), statistical literacy involves understanding and using basic statistics, and the interpretation of these. Statistical reasoning is the way people reason and make sense of statistical ideas and information. Statistical thinking includes knowing how and why particular statistical methods and models are used. These authors compare the three terms to Bloom's taxonomy (Bloom, 1956), with statistical thinking corresponding to the elements of application, analysis and synthesis at the highest levels of the taxonomy. Anderson and Krathwohl (2001) have developed a new taxonomy of the cognitive domain with levels from lowest to highest of remembering, understanding, applying, analysing, evaluating, and creating. Here too statistical thinking might be said to correspond to the three highest levels.

Assessment Methods in Statistical Education Edited by Penelope Bidgood, Neville Hunt and Flavia Jolliffe

Chance (2002) gives a useful summary of various suggestions that have been made as regards defining statistical thinking, too long to reproduce here. She comments that the statistical thinker is able to move beyond what is taught in a statistics course to 'question and investigate the issues and data involved in a specific context'. This is in contrast to statistical literacy and reasoning, both of which can be more narrowly defined.

It is tempting to define statistical thinking as 'what statisticians do', which is something of a tautology. In fact, Gould *et al.* (2006) define statistical thinking broadly as what statisticians do. To some extent this is also the view taken by Wild and Pfannkuch (1999), who researched statistical thinking by interviewing students and statisticians with the aim of finding out about their statistical reasoning. For Garfield *et al.* (2003), 'statistical thinking involves an understanding of why and how statistical investigations are conducted and the "big ideas" that underlie statistical investigations'. Gould *et al.* (2006) state that statistical thinking is statistical literacy made active.

Wild and Pfannkuch (1999) comment that the central element of published definitions of statistical thinking is variation. They themselves divide statistical thinking in empirical enquiry into an investigative cycle, an interrogative cycle, types of thinking and dispositions. Types of thinking fundamental to statistical thinking are: the recognition of the need for data; transnumeration (numeracy transformations made to facilitate understanding); consideration of variation; reasoning with statistical models; and integrating the statistical and contextual.

Although Budgett and Pfannkuch are concerned with assessing statistical literacy (see Chapter 9), they write that they expect their students to think statistically, critically evaluate statements taken from the media and reconstruct them to be statistically sound. They also comment that there is no consensus as regards the definition of statistical literacy. In Chapter 11, Schield, too, is concerned with statistical literacy, but gives some discussion of statistical reasoning and statistical thinking as well as statistical literacy. Katherine Wallman (1993: 1), in her presidential address to the American Statistical Association (ASA), couples statistical literacy with the 'ability to appreciate the contributions that statistical thinking can make in public and private, professional and personal decisions', but does not define statistical thinking per se.

6.2 Teaching statistical thinking

There is general agreement that we should teach what we assess, so in order to assess statistical thinking we should aim to teach it. Watson (1997) claims that the examples she gives can be used for both assessment and teaching. Clearly, teaching statistical methods as such will not necessarily teach statistical thinking, but statistical thinking cannot occur without knowledge of statistical methods. Neither will statistical thinking be learnt by osmosis: guidance and encouragement are needed at all stages, and it needs to be given structure. Chance (2002) writes that statistical thinking might be taught by instilling in students the mental habits and problem-solving skills needed to think statistically and suggests that these include six 'habits', in summary: how to obtain relevant data; reflection on the variables and curiosity about the data and problem; seeing the complete process with each revision; scepticism about the data; relation of the data to the problem context and non-statistical interpretation of the conclusions; and thinking beyond the textbook. (See Chance, 2002, for the full descriptions.) She claims that with the resources that are now available it is viable to instil these habits in students, and expands fully on ways of teaching each of the six 'habits'.

It is possible to teach students the skills associated with statistical thinking, such as communication, critiquing and evaluation through projects of various kinds, but as Wild and Pfannkuch (1999) comment, this is not sufficient in itself. A project does not have to be a substantial piece of work over several months; mini-projects where students have to obtain data, do some analysis, and write about their findings are another possibility. McNiece, in Chapter 10, and Forster and Wild, in Chapter 8, are concerned with communication and how writing can be used to help teach students how to communicate in statistics. Gould *et al.* (2006) feel that the teaching of statistical thinking should involve computer analysis of 'very real data', by which they mean real data that have, for example, missing values or misspelt values.

6.3 The assessment of statistical thinking

Assessment needs to be related to what is taught and how it is taught. The purpose of an assessment and the skills that are being assessed both have to be considered. Given the difficulties in teaching statistical thinking per se, and the links between statistical knowledge, statistical understanding and statistical thinking, assessment of statistical understanding might come through assessment of knowledge and of understanding. As Chance (2002) remarks, much assessment must by necessity rely on questions of a traditional type. Chance gives examples of questions she has used, which to some extent assess the statistical thinking habits that she aims to teach her students. These are to be commended as exemplars of good assessment tasks.

According to Garfield *et al.* (2003), quoting Garfield and Gal (1999), assessing statistical thinking is one of the challenges that need to be addressed. They give two assessment examples involving professionals who want to compare two groups, one where information is given about a confidence interval for a mean with two possible conclusions asking whether these are valid or not, and one involving a stem and leaf diagram where a researcher wants to do the comparison and asks for an appropriate way to approach the problem.

Garfield *et al.*, in Chapter 7, use the words 'critique', 'evaluate' and 'generalise' as those associated with the assessment of statistical thinking. Statistical reasoning, on the other hand, would be assessed by tasks asking for explanations of 'why' and 'how'. They give examples, and there are also some on the ARTIST website (https://app.gen.umn.edu/artist/index.html), which contains other resources also. An account of the ARTIST project (Assessment Resource Tools for Improving Statistical Thinking) is given in Garfield *et al.* (2003).

Many teachers and researchers have written about the assessment of understanding statistics. In Chapter 10, McNiece is concerned with understanding and with developing students' skills in communication. Good communication depends on understanding and this is underpinned by statistical thinking. Part of the assessment that McNiece describes involves the students in writing a report on published examples of epidemiological studies that they have sourced themselves. Jolliffe (2007) contains a section suggesting some similar and other writing assessment tasks as ways of assessing student learning, and points out that marking work such as this is partly qualitative in nature. In assessing learning we also partly assess understanding and thinking.

Forster and Wild too get their students to write about statistics (see Chapter 8). They describe a structured approach to help their students do this, and give an example of an executive summary. Budgett and Pfannkuch, in Chapter 9, say that the assessment examples they give their students are in line with the goals of their course, which include getting students to think statistically. Jolliffe (1997: 197) gives a multiple-choice question from a statistical thinking survey due to Swets *et al.* (1987) that could be used for assessment.

For assessing statistical thinking in the media Watson (1997: 108) suggests a 'three tiered hierarchy: (a) a basic understanding of probabilistic and statistical terminology, (b) an understanding of probabilistic and statistical language and concepts when they are embedded in the context of wider social discussion, and (c) a questioning attitude which can apply more sophisticated concepts to contradict claims made without proper statistical foundation'. The emphasis in the examples Watson gives is to inform teachers as regards instruction and to give students progress reports. These examples are intended more for school-age children than adults, but could be used in any introductory statistics course. In Chapter 9, Budgett and Pfannkuch discuss the use of media examples with undergraduate students.

6.4 Conclusion

There appears to be little that is specifically on teaching or assessing statistical thinking in the literature, but a great deal of relevant material not reviewed here has been published on statistical literacy and understanding. Statistical thinking and understanding are complementary topics and statistical literacy involves both. Thus, assessing understanding or literacy might also be assessing statistical thinking. It is difficult to know whether we are assessing these concepts, since the right answer might be produced for a wrong reason. It could be that oral probing is the only way to be sure. More research is needed in these areas. The paper by Chance (2002) provides a good starting point.

7

Assessing important learning outcomes in introductory tertiary statistics courses

Joan Garfield, Robert delMas and Andrew Zieffler

7.1 Introduction

The introductory tertiary-level statistics course has experienced major changes, both due to changes in the practice of the discipline as well as in what are thought to be the important learning outcomes for students (see Garfield and Ben-Zvi, 2008b). In the 1990s there was an increasingly strong call for statistics education to focus more on statistical literacy, reasoning, and thinking (e.g. Cobb, 1992). One of the main arguments presented is that traditional approaches to teaching statistics focus on skills, procedures, and computations, which do not lead students to reason or think statistically.

The desired result of many introductory statistics courses was consequently re-conceptualised as producing statistically educated students who demonstrate statistical literacy and the ability to reason and think statistically. As a result, there has been more attention paid to distinguishing and defining learning outcomes in introductory statistics courses, in terms of assessing students' statistical literacy, statistical reasoning and statistical thinking (Ben-Zvi and Garfield, 2004b; ASA, 2007).

The shift in emphasis in statistics curriculum and instruction, from developing procedural understanding (i.e. statistical techniques, formulas, computations

Assessment Methods in Statistical Education Edited by Penelope Bidgood, Neville Hunt and Flavia Jolliffe

and procedures), to developing conceptual understanding and statistical literacy, reasoning and thinking must also be mirrored in the assessments used by statistics educators. Following the recommendations of expert psychometricians (e.g. Downing, 2006; Schmeiser and Welch, 2006), assessment development should be preceded by first identifying the desired content domain and the cognitive skill levels to be assessed (test specifications). Educational assessment does not exist in isolation but must be aligned with curricular and instructional goals if it is to support learning (Bass and Glaser, 2004; Mislevy *et al.*, 2003; Wiggins and McTighe, 1998). A clear definition of learning outcomes aligned with assessment provides a blueprint that can be used to identify and develop instruction to support attainment of the curricular goals (National Research Council, 2001).

This chapter illustrates the method of designing and creating an assessment plan that supports learning and that is aligned with the aforementioned goals laid out by the statistics education community.

7.2 Defining learning outcomes

We begin by providing working definitions of the learning outcomes that appear to be valued by the statistics education community, namely, statistical literacy, reasoning and thinking.

Statistical literacy involves understanding and using the basic language and tools of statistics: knowing what basic statistical terms mean, understanding the use of simple statistical symbols, and recognising and being able to interpret different representations of data (Rumsey, 2002).

Statistical reasoning is the way people reason with statistical ideas and make sense of statistical information. Statistical reasoning may involve connecting one concept to another (e.g. centre and spread) or may combine ideas about data and chance. Statistical reasoning involves understanding concepts at a deeper level than literacy, such as understanding why a sampling distribution becomes more normal as the sample size increases (Garfield, 2002).

Statistical thinking involves a higher order of thinking than statistical reasoning. Statistical thinking has been described as the way professional statisticians think (Wild and Pfannkuch, 1999). It includes: knowing how and why to use a particular method, measure, design or statistical model; deep understanding of the theories underlying statistical processes and methods; as well as understanding the constraints and limitations of statistics and statistical inference. Statistical thinking is also about understanding how statistical models are used to simulate random phenomena, understanding how data are produced to estimate probabilities, recognising how, when, and why existing inferential tools can be used, and being able to understand and utilise the context of a problem to plan and evaluate investigations and to draw conclusions (Chance, 2002).

7.3 Assessing statistical literacy, reasoning and thinking

One way to distinguish between these related outcomes is by examining the types of words that are particularly helpful in assessing the different outcomes. Table 7.1 (modified from delMas, 2002) lists words that may be helpful in assessing students' statistical literacy, reasoning and thinking.

Table 7.1 Typical words associated with different assessment items or tasks.

Statistical literacy	Statistical reasoning	Statistical thinking
Identify	Explain why	Critique
Describe	Explain how	Evaluate
Translate		Generalise
Interpret		
Read		
Compute		

For example, to assess student literacy, an assessment might include items such as student interpretations and critiques of selected news articles and media graphs, as well as items pertaining to basic terms and vocabulary. Assessing statistical reasoning may involve items where students are required to respond by 'explaining their reasoning' (e.g. explaining why the standard deviation is greater for one plot of data than for another). Assessments of statistical thinking may include a student project where students pose a problem, collect data, analyse and interpret the results. Next, we present three concrete examples to show how statistical literacy, reasoning, and thinking may be assessed.

7.3.1 Example of an item designed to measure statistical literacy

> A random sample of 30 first year students was selected at a public university to estimate the average score on a mathematics placement test that the state mandates for all freshmen. The average score for the sample was found to be 81.7 with a sample standard deviation of 11.45. Describe to someone who has not studied statistics what the standard deviation tells you about the variability of placement scores for this sample.

This item assesses statistical literacy because it focuses on understanding (knowing) what the term 'standard deviation' means.

7.3.2 Example of an item designed to measure statistical reasoning

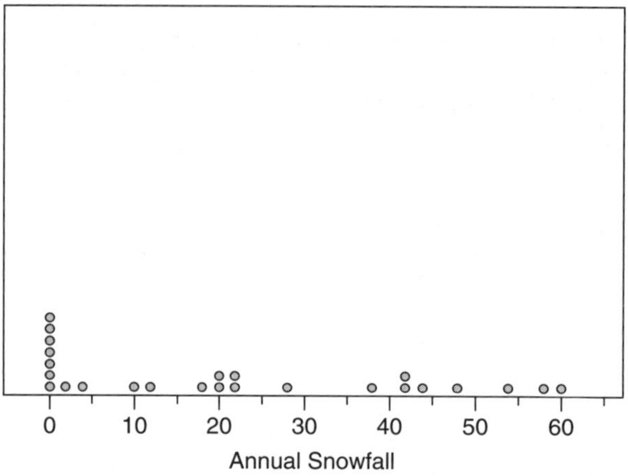

Figure 7.1 The average annual snowfall amounts (in inches) for a random sample of 25 American cities.

> Consider the distribution of annual snowfall amounts for a sample of American cities presented in Figure 7.1. Without doing any calculations, would you expect the mean annual snowfall to be larger, smaller, or about the same as the median? Why?

This item assesses statistical reasoning because students need to reason about how the shape of a distribution affects the relative locations of measures of centre, in this case, reasoning that the mean would be larger than the median because of the positive skew.

7.3.3 Example of an item designed to assess statistical thinking

> A random sample of 30 first-year college students was selected at a public university to estimate the average score on a mathematics placement test that the state mandates for all freshmen. The average score for the sample was found to be 81.7 with a sample standard deviation of 11.45.
>
> A psychology professor at a state college has read the results of the university study. The professor wants to know if students at his college are similar to students at the university with respect to their mathematics placement exam scores. This professor collects information for all 53 first year students enrolled this semester in a large section (321 students) of his 'Introduction to Psychology' course.

Based on this sample, he calculates a 95% confidence interval for the average mathematics placement exam score to be 69.47 to 75.72. Below are two possible conclusions that the psychology professor might draw. For each conclusion, state whether it is valid or invalid. Explain your choice for both statements. Note that it is possible that neither conclusion is valid.

a. The average mathematics placement exam score for first year students at the state college is lower than the average mathematics placement exam score of first year students at the university.

b. The average mathematics placement exam score for the 53 students in this section is lower than the average mathematics placement exam score of first year students at the university.

This item assesses statistical thinking because it asks students to think about the entire process involved in this research study when critiquing and justifying different possible conclusions.

7.4 Comparing statistical literacy, reasoning and thinking to Bloom's taxonomy

These three statistics learning outcomes coincide to a certain degree with Bloom's more general categories of 'knowing', 'comprehending' and 'applying' (Bloom, 1956). We see statistical literacy as consistent with the 'knowing' category, statistical reasoning as consistent with the 'comprehending' category (with perhaps some aspects of application and analysis) and statistical thinking as encompassing many elements of the top three levels of Bloom's taxonomy (application, analysis, and synthesis).

7.5 The role of assessment in curriculum design and evaluation

Wiggins and McTighe (1998) recommend designing a course by beginning with the learning goals (what outcomes are desired) and then working backwards to design the assessments themselves and, lastly, the instructional components of the course. Distinguishing between different types of desired learning outcomes can help statistics educators design assessment tasks that address the different outcomes and create classroom environments that allow multiple instructional methods and assessment opportunities. As a preliminary step in this process, it can often be helpful to develop an assessment blueprint, which requires analysing the purpose and use of each assessment.

7.5.1 The assessment blueprint

In its simplest form, an assessment blueprint is a two-way table where one dimension represents the content domain that is to be assessed and the other dimension represents the cognitive levels that the assessment will cover. The blueprint will help the person writing the test make appropriate judgements about the content that should be sampled from the content domain for the particular assessment being created. An example of an assessment blueprint for an introductory statistics course that primarily focuses on preparing liberal arts students to become informed consumers of statistical information in the world around them is shown in Table 7.2.

In this example, the person creating the assessment has judged that the content dealing with variability is twice as important as the content dealing with shape of a distribution and central tendency. These allocations should reflect the relative emphasis that will be placed on them in the curriculum. However, professional judgement about the relative importance of a topic also needs to play a role. Table 7.2 indicates that the test writer has also judged that half of the test items should be at the cognitive level labelled 'statistical reasoning', with proportionally fewer items at the 'statistical literacy' level, and even fewer items targeting the 'statistical thinking' level.

Table 7.2 Example of an assessment blueprint.

Content	Statistical literacy	Statistical reasoning	Statistical thinking	Total
Central Tendency				20%
Variability				40%
Distribution				20%
Total	30%	50%	20%	

The cognitive level judgements 'must reflect the instructional outcomes and instructional methods used for teaching and learning' (Downing, 2006: 10). The language used to describe both the content and cognitive demand should operate at a level of detail that teachers find useful. It also needs to be of sufficient detail to capture and communicate distinctions among content and levels of cognition.

7.5.2 Example assessment items

After finalising the assessment blueprint, items that match both the content and cognitive level can be selected (or written) so that an appropriate assessment can be created. Several examples are provided in this section.

Example of item that assesses understanding variability at the literacy level

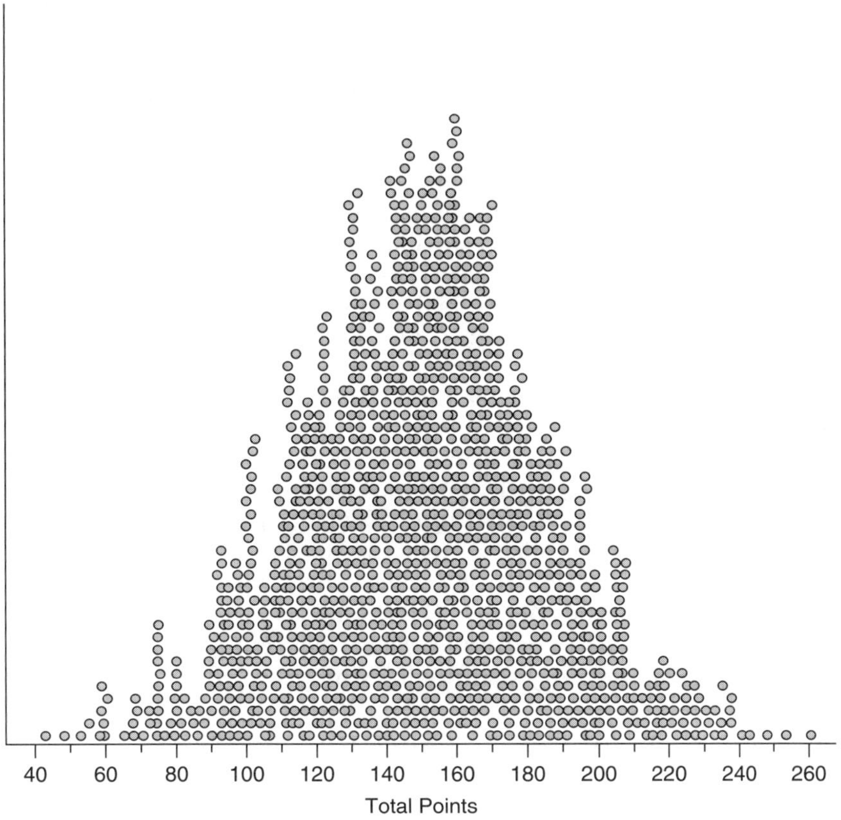

Figure 7.2 Total number of points scored during NCAA play-off basketball games.

a. Given the dot plot in Figure 7.2, estimate the mean and standard deviation, and use them to complete this sentence: The average number of play-off points scored in an NCAA basketball game is about ________, give or take ________.

b. Identify the quartiles and compute the IQR based on the distribution above of NCAA play-off points.

Example of item that assesses understanding variability at the reasoning level

Consider two populations in the same state. Both populations are the same size (22,000). Population 1 consists of all students at the State

University. Population 2 consists of all residents in a small town. Consider the variable Age. Which population would most likely have the largest standard deviation? Explain why.

Example of item that assesses understanding central tendency at the reasoning level

A teacher gives a test to 100 students and determines the mean and median score. After grading the test, the teacher realises that the 10 students with the highest scores did exceptionally well. The teacher decides to award these 10 students a bonus of 5 more marks. Explain how the mean for the new distribution of scores compares to the mean for the old distribution of scores. How about the medians?

Example of item that assesses statistical thinking about distributions of data

A statistics teacher claims that college students spend less time studying than they spend on the Internet, talking on cell phones and other activities. Data were gathered from 120 students in an introductory statistics class about how many minutes they typically spend each week on a variety of activities. Critique the teacher's claim (in two to three paragraphs) by evaluating the boxplots in Figure 7.3 and the histograms and summary statistics in Figure 7.4.

While it may seem fairly easy to create or select items to assess a particular concept at a desired cognitive level, in some cases a single item may assess

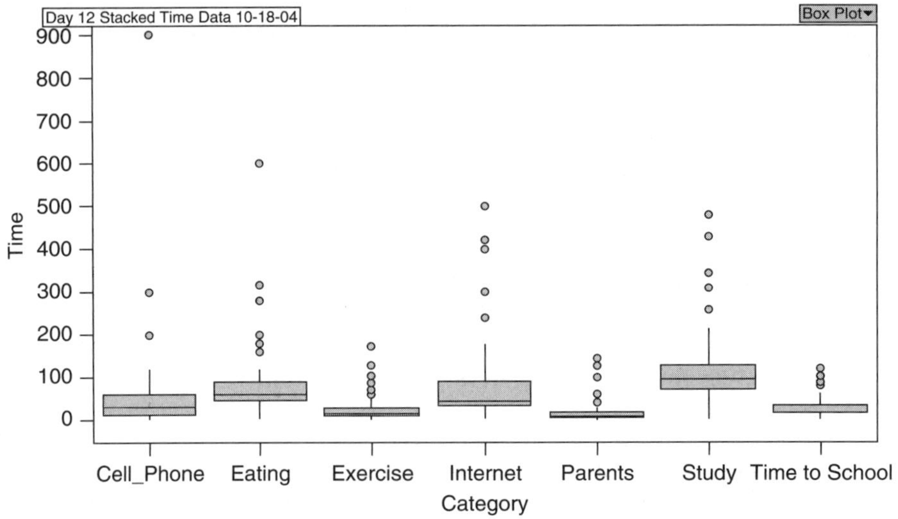

Figure 7.3 Box-and-whisker plots of the time (in minutes) introductory statistics students spend each week on seven activities.

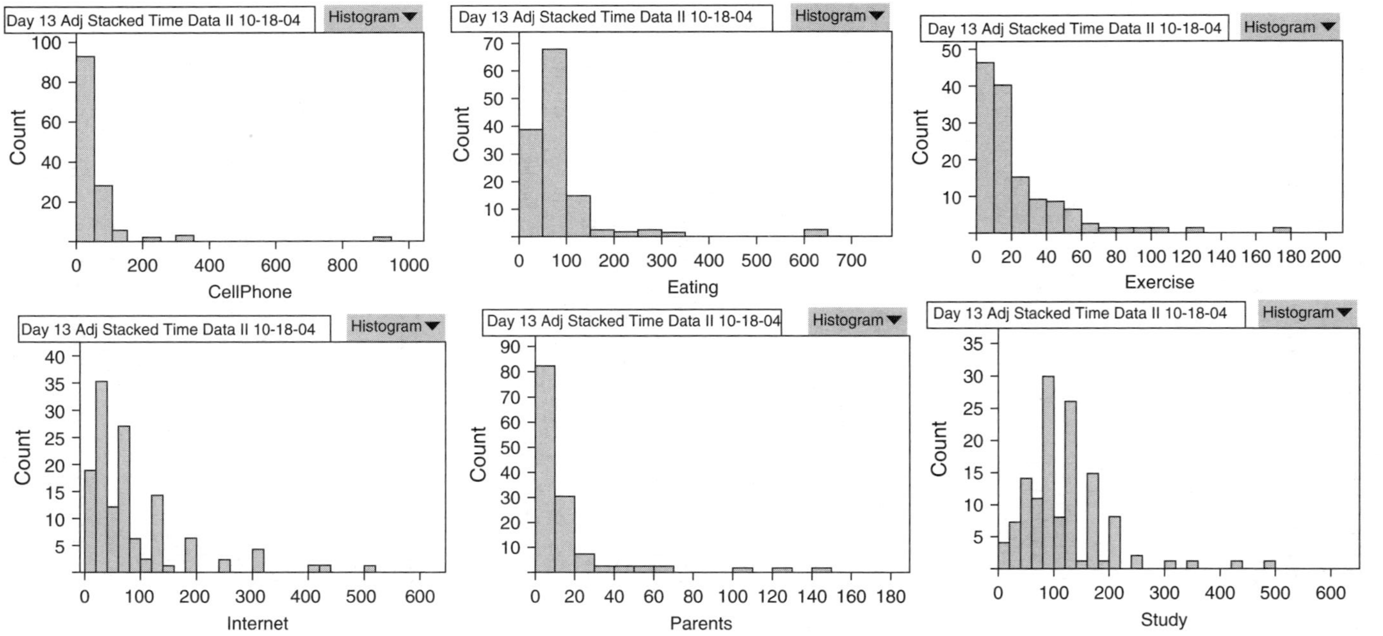

Day 13 Adj Stacked Time Data II 10-18-04 — Summary Table

	CellPhone	Eating	Excercise	Internet	Parents	Study	TimetoSchool
	44.446154	80.122137	22.374046	75.839695	13.717557	115.45802	26.824427
	66.17601	60.499118	25.075681	65.297923	20.910901	73.579602	19.480752
	30	60	14	45	9	94	20
	50	45	20	60	13	60	20

S1 = mean ( )
S2 = s ( )
S3 = median ( )
S4 = xxx ( )

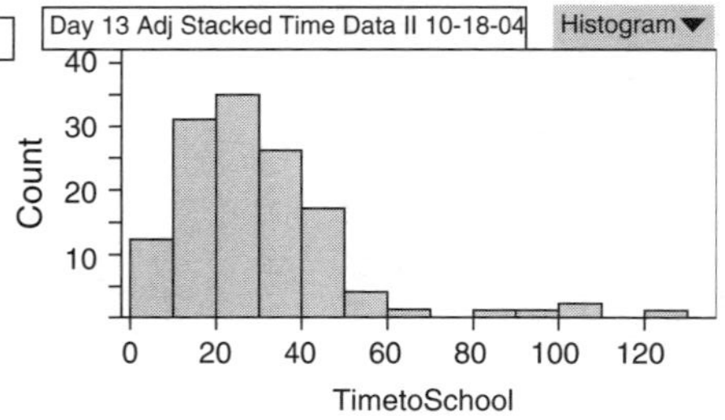

Figure 7.4 Histograms and summary statistics of the time (in minutes) introductory statistics students spend each week on seven activities.

multiple aspects of the content domain (e.g. central tendency and variability). This is often true for items that are meant to assess at the higher cognitive levels such as statistical thinking. We note that in these cases the assessment blueprint can be used as a guide for how marks can be allocated. For example, one item might lead to marks being allocated in more than one area of the content domain.

7.5.3 Student assessment, course evaluation and course revision

Student assessment can also have an important role in evaluating and revising a course. For example, if assessments of statistical reasoning show that students are not able to reason well about important statistical concepts, this may lead to introducing different types of in-class activities that focus on developing this type of reasoning (e.g. having students determine what factors make the standard deviation larger or smaller by using an interactive web applet and discussing their results).

An assessment blueprint for an entire course can be created and used to evaluate the alignment of course goals and student outcomes. This blueprint would ideally classify all the different components of a course's assessment materials and methods, and determine how marks are assigned to the outcomes of statistical literacy, reasoning and thinking as well as to important topics. This process can be helpful in targeting potential mismatches between learning goals, assessment and instruction. For example, finding that although the main focus of instruction and curriculum is at the statistical thinking level, many of the assessment items are actually measuring learning at a literacy level; or that a disproportionate number of assessments items cover topics and material that are covered in just a few weeks of the course. This use of a course assessment blueprint can lead to rewriting assessment items so that they are better aligned with the intended curriculum.

7.6 Online resources for assessing statistical literacy, reasoning and thinking

For the assessment blueprint described above, many of the items and instruments at the (ARTIST) web site (https://app.gen.umn.edu/artist/) can be accessed and modified (see Garfield and delMas, in press). The ARTIST project was funded by the National Science Foundation (ASA-0206571) to develop a web site that provides resources for evaluating students' statistical literacy, reasoning and thinking. These resources were designed to assist faculty who teach statistics across various disciplines (e.g. mathematics, statistics, and psychology), to assess student learning of statistics, to evaluate individual student achievement, to evaluate and improve their courses and to assess the impact of reform-based instructional methods on important learning outcomes.

7.6.1 ARTIST online item database

One of the ARTIST resources, the online item database, contains over 1000 items, keyed to 24 different content areas (e.g. normal distribution, measures of centre, bivariate data), three types of item formats (open-ended, multiple-choice, performance task), and the three types of cognitive outcomes described earlier (statistical literacy, reasoning and thinking). Instructors may search for a subset of items from the database by setting criteria for the item formats, topics, and learning outcomes. A set of linked pages allows the instructor to review, select, and download items into Rich Text Format (RTF) files that may be saved and modified on their own computers with a word-processing program.

7.6.2 The ARTIST topic tests

The ARTIST web site also provides online tests that measure conceptual understanding in 11 important areas that span the content of a first course in statistics. The number of items on each of the topic tests ranges from seven to 15 items, and may be taken online by students either in or outside of class. The items primarily assess statistical literacy and reasoning. The tests can be used to provide formative or summative evaluation information about students' understanding of particular topics, and can be used to review concepts prior to an examination. Instructors can request test reports at any time (typically once all students have completed a test); these are sent to the instructor as email attachments. One report is a spreadsheet that lists the percentage correct and completion time for each student. The other report is a RTF file that indicates the percentage of students who selected each response choice for each item that can be used for review or to provide students with feedback.

7.6.3 The ARTIST CAOS test

The Comprehensive Assessment of Outcomes in a first Statistics course (CAOS) test is a 40-item multiple-choice online test that measures important aspects of statistical literacy and reasoning. Information on the psychometric characteristics of the test is provided in delMas *et al.* (2007). The CAOS test was designed and validated to measure concepts that a group of statistics education experts agreed represent important outcomes in statistical literacy and reasoning that are appropriate and common to most first courses in statistics. This test has been used to measure change in student understanding, for review at the end of a course, as part of the final examination for a course, and for programme evaluation.

7.6.4 Additional ARTIST resources

The ARTIST web site includes many resources, such as references and links to approximately 70 books and articles related to student assessment. There are also many resources on alternative assessment including examples of student projects (with directions to students, scoring rubrics, and samples of student work), as well as other non-traditional assessments (student writing and critiquing activities). Links are provided to information on instruments that may be used to assess student outcomes in a research study, such as measures of achievement in statistics like the Statistical Reasoning Test (Garfield, 2003), and on scales for measuring attitudes and anxiety in statistics students, for example, Survey of Attitudes Towards Statistics (Hilton *et al.*, 2004). All of these resources on the ARTIST web site are freely accessible to instructors of statistics courses.

7.7 Summary

We have presented an argument for an assessment approach that provides detailed information on specific types of learning outcomes in an introductory statistics class. We believe that carefully identifying outcomes and using an assessment blueprint to align assessments to these outcomes is an important part of instruction. We hope statistics instructors will take advantage of the free, high quality assessment resources at the ARTIST web site to evaluate the statistical literacy, reasoning and thinking of their students.

8

Writing about findings: Integrating teaching and assessment

Mike Forster and Chris J. Wild

8.1 Introduction

As Moore (1997) points out, the first principle of assessment is to assess what you value most. In our work in a large second university course in statistics, a course devoted to data analysis, several principles resonated with us.

8.1.1 Motivation

A starting point was the principle that 'the ultimate aim of statistical investigation is learning in the context domain of a real problem' (Wild and Pfannkuch, 1999: 225). Additionally, there was a realisation that to be effective in the real world, our students have to be able to take problems vaguely conceived in natural language terms through the statistical investigation and analysis cycle (Wild and Pfannkuch, 1999, Section 1) to arrive at conclusions that they can successfully communicate to others in natural language. This led us to believe that, for a data analysis course, our ultimate goal should be building in students: the ability to take a set of data with a motivating question; the ability to determine, unaided, what the data is saying; and the ability to communicate that to a non-statistician. From the outset we recognised this as an enormous challenge. Gradually, over time, writing about findings in data became a central focus of both the pedagogy and the assessment in our course.

Assessment Methods in Statistical Education Edited by Penelope Bidgood, Neville Hunt and Flavia Jolliffe

8.1.2 Literature

Written communication underlies many of the assessment methods discussed in Gal and Garfield (1997b). Within that book, the chapter by Watson emphasises the importance of written communication and formative assessment based on critiquing media items, while the chapters by Starkings and Holmes discuss assessing student projects. In their chapter on authentic assessment models, Colvin and Vos include discussion of some high-level scoring rubrics as does the chapter by Keeler. Additionally, the chapter by Jolliffe contains relevant discussion under the heading of open-ended questions.

The major benefits of writing as part of statistical instruction, Radke-Sharpe (1991) states, 'can be summarised in four points: (1) It improves writing skills; (2) it focuses internalisation and conceptualisation of material; (3) it encourages creativity; and (4) it enhances the ability to communicate methods and conclusions'.

Writing 'forces students to go beyond the numerical answers and focus on the meaning of the results' (Parke, 2008), 'to draw meaningful conclusions that connect context and the analysis, and communicate those results to others' (Peck, 2005).

There were a number of calls for an increased emphasis on communication and writing in statistics education in the 1990s (e.g. Hayden, 1989; Radke-Sharpe, 1991; Samsa and Oddone, 1994; Wild, 1994; Watson, 1997). Until recently, there was very little written about quite how we should go about developing and assessing the writing and communication skills of students. The implicit assumption was, in the words of Francis (2005: 1), 'that report writing will come naturally, picked up by a process of osmosis, or that someone else will teach them how to do that – after all, what do mathematicians know about teaching writing skills!' One exception to this general tendency was Stromberg and Ramanathan (1996), who discussed the implementation of writing in an introductory statistics course with some specific writing projects including the keeping of journals and emphasised providing a great deal of instructor feedback.

The two books by Miller (2004, 2005) extensively discuss (professional) writing about statistical findings for a number of audience types, both general and academic. Gal (2003) discusses various formats for written communication from Statistics Agencies. The 2005 International Association for Statistical Education (IASE) Satellite conference on Statistics Education and the Communication of Statistics (Phillips and Weldon, 2005) put a spotlight on statistical communication for statistics educators. Writing about statistical findings is a large and complex area drawing on both statistical expertise and expertise in written communication. Strategies for building statistical writing skills include dedicated writing courses for those who already know a reasonable amount of statistics (Samsa and Oddone, 1994; Weldon, 2007); writing as part of statistical projects (MacGillivray, 2005; Prvan and Ascione, 2005; Biehler, 2007); and writing within mainstream statistics courses.

Samsa and Oddone (1994: 118) describe how they came to conclude, from their experience of teaching a dedicated writing course, that, 'writing instruction and assignments should instead be more broadly dispersed among multiple courses in the curriculum'. Parke (2008) discusses incorporating statistical communication in graduate education with a view to writing research articles. Francis (2005), Lipson and Kokonis (2005), and Peck (2005) discuss writing in university introductory statistics courses. Francis particularly emphasises the subtleties of language. Forster *et al.* (2005) and Forster and Smith (2007) provide earlier accounts of our own work. Additionally, in a move that we applaud, some introductory texts such as Peck *et al.* (2008) are beginning to incorporate in almost every chapter a standard section on interpreting and communicating the results of statistical analyses.

8.1.3 Goals

Assessment rewards achievement – something that has been done in the past. The real goal of education, however, is to build capabilities within the students themselves that will facilitate achievement in the future. The course that is at the centre of this chapter is just one of a sequence of applied statistics courses. In thinking about the education of statisticians who will be useful practitioners in the real world we should think in terms of the whole sequence and what students emerging from the sequence should be capable of doing by the end of it. Ideally, each course in the sequence will:

- Increase each student's *technical* capability (traditionally 'the content').
- Increase each student's *integrative* capability (ability to interlink all they have learned so far).
- Increase each student's *recognitive* capability (for recognising where their tools are likely to prove useful; see Wild (2007, Section 3).
- Increase each student's *distillatory* capability (for distilling information and extracting meaning).
- Increase each student's *communicative* capability.

In other words, each course should take what exists of every one of these capabilities from prior courses and further develop them. Teaching strategies are required in order to accomplish each of these goals. Almost all pedagogical attention in statistics has been focused on 'content', the first and easiest item (and even then, often degenerating to 'teach more topics'). By stressing writing, we can work simultaneously on the capabilities for integrating, distilling information and communicating.

The underlying validity of the mantra of assessment, 'assess what you value', is unassailable. It does, however, have a flip-side: 'teach what you assess'. In the race to make assessment mirror authentic statistical activities, many of us fall into the trap of assessing things we do not teach. This is particularly true of the ability to write and present findings or make an argument. Unless the students' capabilities are developed in these areas, we are simply giving credit for pre-existing skills. This is fine for forms of a credentialing that attempt to predict future capabilities, but in terms of measuring 'what I learned on this course' it is grossly unfair. The message behind the mantras – 'assess what you value' and 'teach what you assess' – is to strive to pursue the goals even though you may know you are falling well short of them.

8.2 Background

We now describe the background to the writing initiative in our course starting with a discussion of what our students have experienced in the prior course, then moving to the makeup of our student body, the nature of the course, the history of the writing initiative and the forms of assessment used in our course.

8.2.1 The first course

The first-year introductory course at the University of Auckland is delivered efficiently to very large numbers – approximately 4500 per year. The main summative assessment vehicle is a multiple-choice examination. Wild *et al.* (1997) give a variety of strategies for encouraging thinking and deeper learning using multiple-choice. No matter how clever you are with multiple-choice, however, it comes down to recognition that items in a list are right or wrong, promising or implausible; it is about recognising something from a list presented by someone else rather than having something bubble up from the student's own mental resources. At the University of Auckland we want students to develop the ability to put together entire sequences of initiatives, activities and ideas, unprompted.

A start is made in first-year course assignment work, heavily scaffolded by step-by-step prompting. In terms of communication, students only have to come up with a couple of short sentences at a time, explaining an idea, or interpreting such things as the results of significance tests and confidence intervals in context. This is not a bad thing. All courses have to make some hard trade-offs between competing priorities. We know that trying to do everything at once is a sure-fire way to accomplish nothing. But in the second course we want to build upon the foundations laid in the first course and move the students to a new level of self-sufficiency at which they can operate with much less scaffolding to aid them.

8.2.2 Our course and its students

Our course is a second course in statistical data analysis. It is a large service course, taken by approximately 1300 students per year (accumulated over two semesters and a summer school). In both formal and practical terms, its biggest client is the undergraduate programme in marketing. It is also required for a number of graduate programmes in business and economics. It is also taken voluntarily by large numbers of students who intend to major in experimental sciences, including psychology and mathematics. It is the main prerequisite for all third-year courses in applied statistics – the majority of those who become statistics majors do not decide to do so until the end of this course because they started their degrees planning to graduate with some other major. The course also figures as a nearly compulsory component of several small cross-disciplinary science programmes via degree specifications such as 'choose two courses from this list of three'.

The students are therefore diverse in terms of their academic interests, in the extent to which they have had to write in the past, and in terms of cultural and language backgrounds. In our 2008 second semester class, we had students who were born in around 50 different countries with only about 50% having English as their first language: this is the norm rather than the exception. Such diversity raises both equity and retention issues when we want to make writing a prominent aspect of the course. Many non-native speakers of English often choose courses from the mathematical sciences in order to avoid high language demands ('language flight'). The students of this second course do, however, have a common background in statistics, as virtually all have taken essentially the same first-year introductory course.

Our course is a practical computer-based course in data analysis using the R package (R Development Core Team, 2008). It does not therefore, pay a great deal of attention to up-front aspects of the statistical enquiry cycle (e.g. question formulation, choice and definition of variables, and formulation of a sampling or experimental design; see Wild and Pfannkuch, 1999, Section 1) except insofar as they provide a context for the analyses to be performed and their resultant findings. The central content consists of gaining familiarity with several classes of models and their use in data analysis, and with extracting and communicating the findings from these analyses. The classes of models worked with are multiple regression and analysis of variance (and their simpler special cases), chi-square tests, odds ratios, logistic regression, and an introduction to time series models. The course is data-rich: there are over 40 analysis Case Studies (an example is given in Section 8.5). The assessment instruments alone require another 26 data sets per semester.

We want our students to take away from such a course the ability to:

- Recognise what types of analysis are likely to be fruitful for a given data set.
- Analyse the data successfully using a modern computer analysis package.
- Abstract what is informative and important from package output.
- Communicate these findings to others.
- They should not merely be able to communicate their findings to the statistically trained, but also to those who have no formal understanding of statistics or its terminology. This chapter concentrates on the abstraction and communication steps.

8.2.3 History

In 1997, we began asking students in examinations to write paragraphs describing the main findings from an analysis of data. We asked that they write in non-technical language that would be easily understood by those who have no formal statistical training. We called such paragraphs 'Executive Summaries' and the name, which has stuck even as the concept has developed, conveys our intent extremely well, particularly if we think in terms of a consultant working for a client who asks: 'Now that you have analysed my data, what's the bottom line? What have you learned from the data that I need to know?' Until we get data analysis students to this point, they have not actually learned anything that is particularly useful.

Initially, students were only provided with a small number of examples written by the teachers intuitively, without any clear idea of how they were going about it. We were troubled by the terrible quality of much of the writing we received in response and began to experience qualms about the efficacy and the ethics of assessing skills we were not actually teaching. The realisation dawned that at this stage of their development most students need a much more systematic approach, with a great deal more scaffolding and practice, to write successfully. The highly systemised forms of writing we are now using in our course and the teaching processes around them, are the culmination of a ten-year learning process. Francis (2005: 1) had very similar experiences and came to the same conclusion: 'Report writing needs to be taught explicitly, and in the context of understanding what you are trying to convey to your audience'.

8.2.4 Assessment

In our second course, more than half of the assessment takes the form of writing reports of findings from data analyses. There are two types of reports. Executive Summaries are bottom-line, what-we-have-learned reports written for a non-technical audience. The ability to do this is one of the main aims of the course. 'Technical Notes' are basically personal journals of what has been noticed during a data-analysis journey. They provide the raw materials from which the Executive Summary is to be abstracted. The scaffolding given is a general pattern for these types of report. These patterns, or scaffolds, are intended to be internalised as the course progresses. They are reinforced constantly in lectures, tutorials and assignments, but are withdrawn totally by the final examination; by which time students have to remember the scaffolds. The precise nature and rationale of Executive Summaries and Technical Reports, strategies employed, and the integration of teaching and assessment, are described in Section 8.3.

For their assignment work, the students only get a description of the data (cf. Section 8.5.1), the names of the variables and the data file. From these basic ingredients and the examples they have seen in class, they are required to analyse the data and write up reports of their findings. We put our students in the same situation that we face as statisticians. No step-by-step instructions are given.

They are required to choose the appropriate type of analysis to perform, analyse the data on the computer and report their findings in their Technical Notes and Executive Summary. We also require them to attach the relevant computer output they have generated for each analysis as an appendix to their assignment. The computer output is worth no marks but is included to enable us to determine if they have made any errors in the analysis phase (e.g. not transforming their data when necessary) and give partial credit where it is due. Choosing the types of analysis to use is easier in assignments than the final examination because the students can (and do) expect to be making heaviest use of the tools that they have learned about most recently.

Assignment work is both formative and summative, and counts for 20% of their final grade. The course has a multiple-choice mid-semester test, which counts for 20% of the final grade and a final examination which counts for 60%. The majority of the mid-semester test focuses on interpretation of data-analysis output presented in a substantial appendix of data descriptions and computer output known as 'the data appendix'. Such questions are complemented by a smaller set of what-to-use-where questions and questions about concepts.

The final examination also includes an even longer data appendix. The first section of the final examination, counting for 30% of the marks, is comprised of multiple-choice questions, including interpretation of some of the computer output, predominantly on analyses from the end of the course (time series and logistic regression) and with the remainder being what-to-use-where, theory and concept questions. Another 20% consists of short answers, including calculations (odds ratios and fitted values in regression) and identifying the appropriate form of analysis to use. The final 50% requires the students to write Technical Notes and Executive Summaries, unaided, on analyses reported in the data appendix. We generally have our students write two sets of Technical Notes and four Executive Summaries.

8.3 Technical notes and executive summaries

We start with an overview of the goals for and nature of the systemised forms of writing used in our course, namely Technical Notes and Executive Summaries. We then give detailed prescriptions for both types of document. Examples are given in Section 8.5.

8.3.1 Overview of technical notes and executive summaries

In consulting reports, Executive Summaries (the bottom line of what has been learned) are backed up by fuller expositions and appendices, which really just address any real-life client's requirement: 'Give me some reasons for trusting what you say'. Isolating and communicating the bottom line is a very tall order, demanding an understanding of many dimensions of a problem, synthesis and integration, and selective judgement (what is bottom-line versus what is secondary versus what should be ignored altogether). There are many audiences for writing, with many different stylistic conventions, but the common denominator for communicating is working out, ‘What is the story that I have to tell?’ This is the central prerequisite that needs to be fulfilled before we start thinking about how best to tell a story to a particular audience in a way that its members are likely to be receptive to and understand.

We have not tried to proceed as far as formal academic writing or journalism in this course. If through writing Executive Summaries our students learn to write clear abstracts and communicate the bottom line in a way that can be widely understood, we have laid an excellent foundation for their further education. The discipline of having to write for a general audience also brings students to a deeper understanding both of statistics and the results of their analyses: the process demands that they wrestle actively with the real-world meaning of concepts rather than hiding behind technical jargon.

Our Executive Summaries and Technical Reports are disciplined forms of writing through which students can learn to:

- Abstract the cogent features of an analysis.
- Improve their ability to communicate these findings coherently.

In the process of doing this, our practice is to:

- Repeatedly reinforce particular elements of ‘What should I be paying attention to?’ in computer output and in writing about it.
- Repeatedly reinforce basic principles of statistics and the data analysis cycle.
- Gradually improve the student’s understanding of the nature of data, of models and what they can tell us.

While the ability to successfully write Executive Summaries is one of our goals, the more recently developed Technical Notes are a means towards that end. Experience showed that, to make the abstraction and writing processes easier for students, we needed to add a stage intermediate between looking at the output and writing a summary. The Technical Notes provide a concise set of information, abstracted from the computer output during a ‘noticing’ phase, upon which the Executive Summary can then be constructed. In addition they are intended to deliver on the second set of (three) bullets above. Since the Technical Notes do employ technical language, they should also help prepare students to write scientific papers that have statistical analyses in them.

We believed it important that students have structures for both Technical Notes and Executive Summaries that were simple enough to internalise so that they could write without needing to consult a checklist. To do this across a wide variety of types of analysis meant coming up with a structure that could be applied to them all with a minimum of analysis-specific local variations. We have accomplished this with a simple basic shell that is common to all Executive Summaries (and another that is common to all sets of Technical Notes), and then emphasise the few specific features that require special consideration for a particular type of analysis.

The Technical Notes and Executive Summaries were originally developed to help students write coherently and insightfully in their assessments. Subsequently they became core parts of the medium of instruction – the integration of teaching and assessment referred to in the title of this chapter. The primary way that data is interacted with in lectures and tutorials is by taking some of the approximately 40 Case Studies and going through an analysis performed live on the computer, accompanied by a staged 'noticing' phase to produce Technical Notes, which are then translated into an Executive Summary. The classroom experience thus models what is expected in assessment. Students' ability to write the reports is built up through all of the assignment work, where some support is available, and then finally they have to do it entirely on their own in the examination.

Section 8.5 contains a case study from the course complete with its data description and R output (8.5.1), a Technical Notes exemplar (8.5.2), and an Executive Summary exemplar (8.5.3). The case study involves data collected at Auckland Zoo to assess the effectiveness of a television advertising campaign in increasing attendance figures. Factors taken into account include weather, a time trend and a day-type effect (weekday, weekend day or public holiday).

8.3.2 Technical notes

An example of a set of Technical Notes is given in Section 8.5.2. Technical Notes are a record of the 'noticing' phase, 'of a statistician for a statistician' written using technical language in a way that is designed to reinforce processes involved in the Data Analysis Cycle depicted in Figure 8.1.

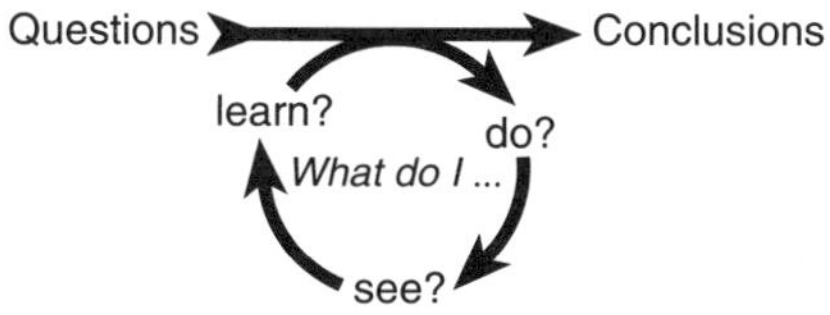

Figure 8.1 Data analysis cycle.

Technical Notes are written under three headings: Exploratory Analysis, Checking Assumptions, and Statistical Inference.

Exploratory analysis Students report the main features they see in an appropriate plot that they have generated of their data. Depending on the data and analysis technique that is required, we expect one or two main features that stand out in their plot to be discussed.

Checking assumptions A report, in formal statistical language, the hypotheses associated with and the results of any tests (e.g. Shapiro-Wilk test, Levene test) the students have performed, including P-values. They are also required to comment on any plots they may have used (e.g. Normal Q-Q plot). They must discuss any transformations of the data they perform and any model building steps that are required (e.g. building a regression model using Backward Elimination).

Statistical inference A brief report on the main statistical findings from the students' final model, once again in formal statistical language. We generally restrict this section to a discussion of the results of the formal test(s) that were performed (t-test, F-test, Chi-square test), but we do not require our students to report estimates of the size of any effects (confidence intervals) in this section. This could, of course be done here but we emphasise the reporting of confidence intervals in the Executive Summary and do not want them to have to do it twice.

8.3.3 Executive Summaries

An Executive Summary summarises the big messages found in the data for a non-technical audience. Students are penalised for including any technical jargon. In rare exceptions, where technical language seems unavoidable, (e.g. discussing interaction in two-way ANOVA) students must include a non-technical description of any technical terminology used, couched in the context of their data. An example of an Executive Summary is given in Section 8.5.3.

The Executive Summary has five paragraphs: Introduction; A Catch-all; Strength of Evidence; Quantification; Summary.

Introduction A one- or two-sentence description of the data and the purpose of the analysis.

A catch-all A 'catch-all' paragraph providing any important information the reader of the report needs to have in order to understand why and how the results are reported. This is not needed for all analyses but is the place where any local analysis-specific variants are discussed. If the data required a transformation before the final analysis was performed, the students need to state this and any implications the transformation has for the interpretation of their findings (e.g. multiplicative, medians rather than means). In two-way ANOVA, they are required to discuss whether the factors interact and give a non-technical description of what the interaction, or lack thereof, means in the context of their data. (This forces them to come to grips with a concept they often find difficult.) For regression, we require the students to discuss the fit of their model and whether it is useful for prediction.

Strength of evidence Students report the strength of evidence for any effects they have detected, based on the P-values, as well as stating which effects, if any, are not significant.

Quantification Quantification means a discussion of the sizes of any significant effects by interpreting the confidence intervals generated in the analysis phase (the reason for not asking them to quantify non-significant effects is simply to prevent the reports getting too long and tedious).

Summary Finally, if the report is quite long or the main message our students see in the analysis gets lost in a mass of figures, a one or two sentence summary of the major findings is required. We have found that a very brief summary of their main findings is often useful for analyses of two-way tables of counts, multiple regressions and any data that have come from a designed experiment.

8.4 Discussion

A standard education strategy for anything which at the outset is indigestibly complex, is first to establish an oversimplified version and then repeatedly cycle back adding complexity in digestibly small chunks until sufficient realism has been incorporated to serve a real-world purpose. Rather than being a final destination, the forms of writing we establish in our course are the early stages of a development process. Subsequently, increased flexibility in the way students write should be encouraged and scaffolding progressively removed as they develop their own judgement about the nature of the messages to be conveyed and the target audience. But judgement is a product of experience. It does not come from being thrown overboard without a life-preserver.

There can be a degree of clumsiness in the reports. As teachers we can often see how to get to the bottom line more quickly. Sometimes the framework encourages students to comment on things that a professional might not quite classify as "bottom line". These are unfortunate consequences of the compromises made in trying to have one teachable pattern that covers basically everything in the course. Seeing the minor defects in the writing and knowing how to address them may feel intuitive to us as educators, but it is really the product of years of experience – experience that the students do not have.

Our structured approach to teaching the communication of the main findings from analyses of data has made a dramatic difference to the coherence and coverage of the reports on data that we get from the bulk of students. They now generally find the patterns easy to learn and can write passages that capture most of the main features of the data and actually make sense! The fact that this report-writing repeatedly forces students to integrate information and form a big picture helps counteract the fractionation of knowledge we educators so often induce by spending so much of our teaching time spelling out details.

The major weaknesses of our approach are flip-sides of its strengths. The reports we get our students to write are very prescriptive in what we get them to comment on and highly structured in their design. While this has made it easier for our students (especially the second language students), and made grading work relatively easy, it can stifle creativity in the report writing. So, where to from here? We are working with others on the transition between writing models used in this second course and writing models in the array of third-year applied courses that follow on from it. There are features that have been very successful elsewhere, but may be difficult to incorporate into courses with large numbers of students. For example, having students discuss and critique one another's writing and using peer assessment are reported to have been valuable learning devices by Parke (2008) and Stromberg and Ramanathan (1996). Currently our course focuses almost solely on the data analysis phase of the investigative cycle. There are plans afoot to introduce own-choice, group projects (cf. MacGillivray, 2002) so that students experience the whole cycle at least once. Students will then also have to learn to write project reports.

Teaching our students to communicate their findings has, on balance, been a very successful aspect of our courses. We believe it has assisted our students to appreciate that an analysis of data is not completed when the computer analysis is done but that the contextual interpretation of the findings and their communication are equally important. We have managed to tailor our report writing in such a way that it reinforces the data analysis cycle and the Technical Notes have assisted our students in recognising and selecting for comment only the relevant output from the computer analysis.

Students generally appreciate that their writing skills are being improved. In an evaluation conducted at the end of last semester 71% agreed, 'The course improved my skills in written communication', while 92% agreed with the statement, 'I have found the case study based learning very helpful'. It was particularly pleasing to get an unsolicited email from the top student in the class, saying, 'This course has enabled me to improve my analytical and writing skills'.

8.5 Zoo data case study

These data represent 455 days of attendance figures at Auckland Zoo, beginning on 1 January 1993 (although some data points have been omitted due to missing values). There are 440 complete observations. It was of interest to assess whether an advertising campaign was effective in increasing attendance.

8.5.1 Data description

The variables measured were:

- attendance: number of visitors
- time: time in days since the start of the study (1/1/93)
- sun.yesterday: hours of sunshine the previous day
- tv.ads: average daily spend on TV advertising in the previous week (in $000's)

- nice.day: assessment based on number of hours of sunshine: 1 = Yes, 0 = No
- day.type: 1 = ordinary weekday, 2 = weekend day, 3 = school holiday weekday, 4 = public holiday

The computer output is given in Figures 8.2 and 8.3.

8.5.2 Technical notes

Exploratory analysis:

- The pairs plot reveals little, except a possible increasing trend in attendance over time.
- The dot plot of attendance by whether it was a nice day shows attendance is higher, on average, on nice days.
- The dot plot of attendance by day type shows that attendance is higher, on average, on weekends, school holiday weekdays and public holidays.

Checking assumptions:

- The observations appear to be independent.
- The residual plot for the linear model shows evidence of heteroscedasticity. The residuals are more spread out for weekends and school holiday weekdays.
- After fitting a model for log attendance, the residual plot shows random scatter about 0, but there are clusters of points by type of day.
- The plot of Cook's distance shows no unusual observations.
- The Normal Q-Q plot shows that a line through the points has the characteristic shape of a left skewed distribution.
- The Shapiro-Wilk test provides extremely strong evidence against the hypothesis that the errors from the log model have a normal distribution (P-value = 7.111×10^{-16}). However, we can rely on the Central Limit Theorem as the sample size is 440.

Statistical inference:

- The F-test for regression provides very strong evidence against the hypothesis that none of the variables are related to the response (P-value ≈ 0).
- The Multiple R^2 is 0.7138 indicating that 71% of the variation in log attendance is explained by the variation in time, the amount of sun the

previous day, the average amount spent on TV ads per day in the previous week, whether it was a nice day and the day type, so the model will not be very useful for prediction. Prediction will also be unreliable as there are problems with normality.

```
> pairs.20x(data.frame(attendance,time,sun.yesterday, tv.ads))
```

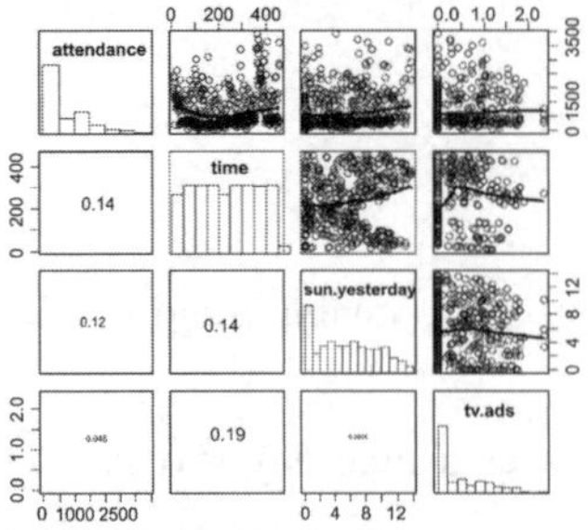


```
> stripchart(attendance~nice.day,vert=T,pch=1,
     method="jitter",main="Attendance by nice day")
```

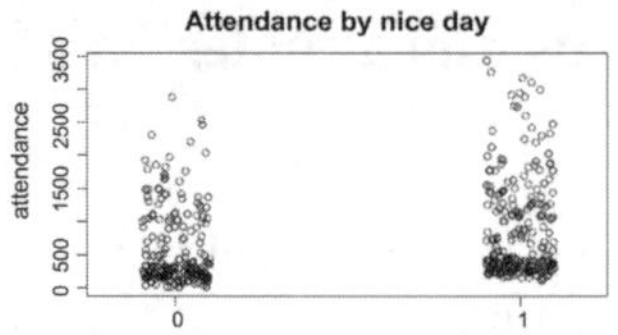


```
> stripchart(attendance~day.type,vert=T,pch=1,
     method="jitter",main="Attendance by day type")
```

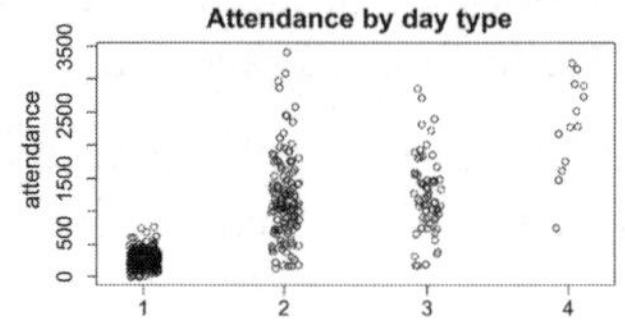


```
> day.type1.<-factor(day.type)
> zoo.fit<-lm(attendance~time+sun.yesterday+tv.ads+
                       nice.day+day.type)
> plot(zoo.fit,which=1)
```

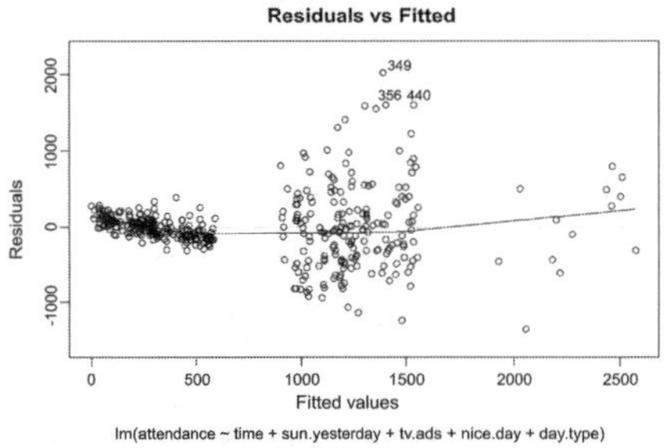


```
> plot(residuals(zoo.fit)~fitted(zoo.fit),
   type="n",main="Residual plot (fitted) by day type")
> text(fitted(zoo.fit),residuals(zoo.fit),day.type)
```

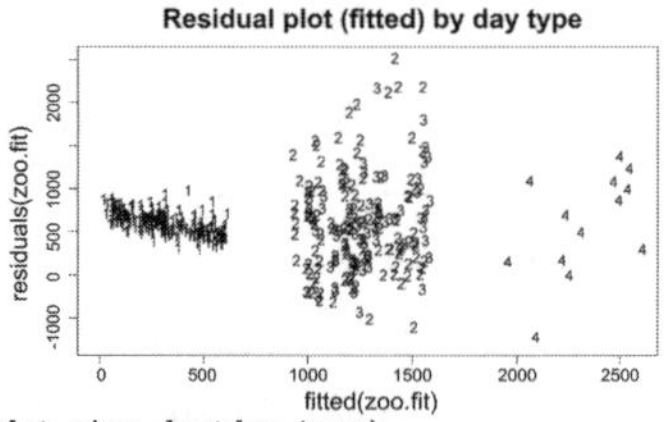


```
> zoo.fit1<-lm(log(attendance)~time+sun.yesterday+tv.ads+ nice.day+day.type)

> plot(residuals(zoo.fit1)~fitted(zoo.fit1),type="n", main="Residual plot (fitted) by day type")
> text(fitted(zoo.fit1),residuals(zoo.fit1),day.type)
```

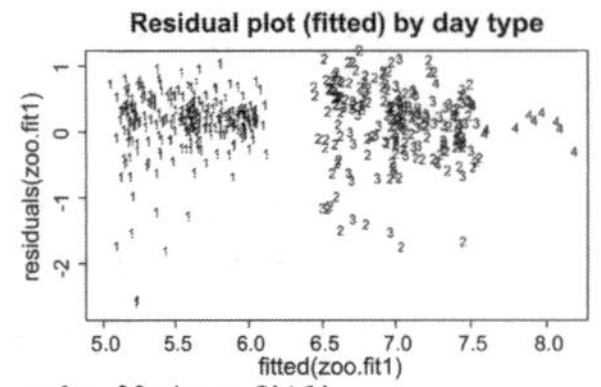


```
> cooks.20x(zoo.fit1)
```

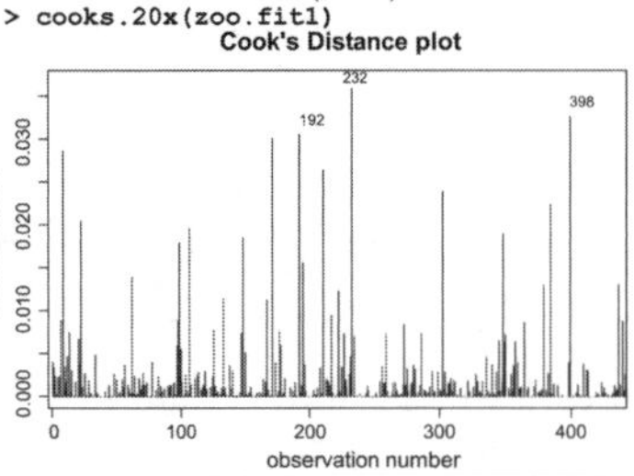


```
> summary(zoo.fit1)

Call:
lm(formula = log(attendance) ~ time + sun.yesterday +
                    tv.ads   + nice.day + day.type)
 .....
 Coefficients:
              Estimate Std.Error t value Pr(>|t|)
(Intercept)     4.940   0.0650   75.95   < 2e-16 ***
time            0.001   0.0001    4.980  9.21e-07 ***
sun.yesterday   0.015   0.0065    2.387  0.01740 *
tv.ads          0.124   0.0428    2.915  0.00374 **
nice.day        0.433   0.0515    8.408  6.66e-16 ***
day.type2       1.399   0.0561   24.928  < 2e-16 ***
day.type3       1.492   0.0736   20.271  < 2e-16 ***
day.type4       2.062   0.1454   14.180  < 2e-16 ***
---
Signif. codes: 0`***'0.001`**'0.01`*'0.05`.'0.1` '1

Residual standard error: 0.5097 on 432 degrees
                                        of freedom
Multiple R-Squared:0.713, Adjusted R-squared: 0.709
F-statistic: 153.9 on 7 and 432 DF,  p-value: 0
```

Figure 8.2 Computer output, Part I.

```
> plot(zoo.fit1,which=2)
```

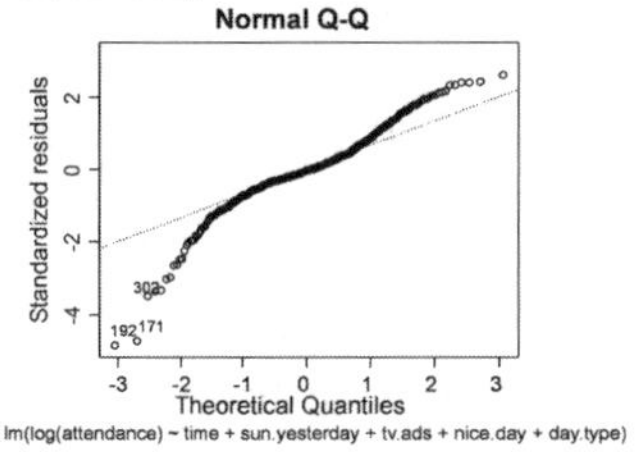


```
> shapiro.test(residuals(zoo.fit1))

        Shapiro-Wilk normality test

data:  residuals(zoo.fit1)
W = 0.9058, p-value = 7.111e-16
```

```
> ci.reg(zoo.fit1)

                95 % C.I.lower 95 % C.I.upper
(Intercept)            4.81295        5.06864
time                   0.00058        0.00135
sun.yesterday          0.00277        0.02861
tv.ads                 0.04069        0.20906
nice.day               0.33238        0.53518
day.type2              1.28909        1.50977
day.type3              1.34816        1.63766
day.type4              1.77691        2.34878
```

```
> exp(ci.reg(zoo.fit1))

                95 % C.I.lower 95 % C.I.upper
(Intercept)         123.094210     158.957997
time                  1.000580       1.001351
sun.yesterday         1.002774       1.029023
tv.ads                1.041529       1.232519
nice.day              1.394283       1.707756
day.type2             3.629482       4.525690
day.type3             3.850334       5.143121
day.type4             5.911561      10.472785
```

Figure 8.3 Computer output, Part II.

- The intercept is not meaningful.
- We have very strong evidence against the hypothesis that the slope coefficient associated with time is equal to 0 ($\mathrm{P}-\text{value} = 9.21 \times 10^{-7}$).
- We have evidence against the hypothesis that the slope coefficient associated with the amount of sunshine the previous day is 0 (P-value = 0.0174).
- We have strong evidence against the hypothesis that the slope coefficient associated with the amount spent on TV adverts per day in the previous week is 0 (P-value = 0.00374).
- We have extremely strong evidence against the hypothesis that there is no difference in attendance if it was a nice day compared to when it is not ($\mathrm{P}-\text{value} = 6.66 \times 10^{-16}$).
- We have extremely strong evidence against the hypotheses that there is no difference in attendance between the weekend and an ordinary weekday, a school holiday weekend and an ordinary weekday, or a public holiday and an ordinary weekday (P-values ≈ 0).

8.5.3 Executive Summary

These data were collected to assess whether a TV advertising campaign was successful in increasing the number of visitors to Auckland Zoo.

We had to transform the data; as a result our statements are about the effects on median attendance, are multiplicative in nature and will be expressed in terms of percentage changes.

The model explained 71% of the variation in attendance figures and so would not be very good for predicting the number of visitors to Auckland Zoo.

We have strong evidence an increase in the average daily amount spent on television advertising in the previous week is associated with an increase in median attendance.

We have very strong evidence of a time trend in median attendance, with median attendance increasing slowly over time.

We have some evidence that each additional hour of sunshine on the previous day is associated with an increase in median attendance.

We have very strong evidence that median attendance increases if it is a nice day.

We also have extremely strong evidence that median attendance is affected by the type of day, with Public Holidays, weekends and school holiday weekdays having higher median attendance than ordinary weekdays.

Holding everything else constant, we estimate that:

- Each additional \$1000 spent on television advertising per day during the previous week is associated with an increase in median attendance of between 4% and 23%.
- Median attendance increases by between 6.0% and 14.4% every 100 days.
- For each additional hour of sunshine on the previous day, median attendance increases by between 0.3% and 2.9%.
- For days that have enough sunshine to be categorised as a 'nice day', median attendance is between 39% and 71% higher than if there is insufficient sunshine to categorise the day as a 'nice day'.
- Median attendance is higher by between 263% and 353% if it is a Saturday or Sunday than if it is an ordinary weekday. Median attendance is higher by between 285% and 414% if the day is a school holiday weekday than an ordinary weekday. Median attendance is higher by between 491% and 947% if the day is a Public Holiday than if it is an ordinary weekday.

The advertising campaign appeared to be successful.

9

Assessing students' statistical literacy

Stephanie Budgett and Maxine Pfannkuch

9.1 Introduction

Statistical information is prolific in the media but the citizen without statistical literacy may be misled or have difficulty in interpreting and challenging statements such as the following:

- Obesity in children doubles in 11 years (*NZ Herald*, 23/03/2004).
- Smoking ban sees coronaries halved (*NZ Herald*, 06/04/2004).
- Labour has its nose just ahead by 44.3% to National's 43% in the 18–24 age group (*NZ Herald*, 29/03/2008).
- Herceptin patients had a 32% lower risk of dying than others (*NZ Herald*, 29/07/2006)
- Experts have blamed 'mad January' for the 10 murders since the start of the year – with the finger being pointed to everything from Christmas stress to alcohol to a full moon. (*NZ Herald*, 31/01/2008)
- One of the most detailed studies ever made of the contested health phenomenon of deep vein thrombosis, has found it may occur in as many as one in every 100 frequent long-haul air travellers. That means that for every jumbo jet landing here from a long-distance flight, around three or four passengers will suffer a potentially fatal blood clot (TV One News, 19/12/2003).

Assessment Methods in Statistical Education Edited by Penelope Bidgood, Neville Hunt and Flavia Jolliffe

Students on our undergraduate course at the University of Auckland entitled 'Lies, Damned Lies, and Statistics' are expected to able to 'think statistically' and critically evaluate such statements, and also to reconstruct them to be statistically sound. They need to check how obesity is defined; judge whether the lower coronary rate was caused by the smoking ban; consider the margin of error for a difference and verify the claim that Labour is ahead; comprehend the difference between risk, reduced risk and relative risk; be aware of Poisson clumping; and have a sense of a probability distribution of outcomes. Hence, our chapter aims to describe the methods and challenges we face in assessing students' statistical literacy.

In Section 9.2, we briefly describe the course and in Section 9.3, we discuss our assessment framework. In Section 9.4, we describe how our assessment tasks assess statistical literacy. Some exemplars of students' responses to assignments as well as examples of questions from tests and examinations are given. In Section 9.5, we present a qualitative analysis of interviews conducted with past students in an effort to determine the impact of the course on developing their capacity to use the statistical thinking skills learnt in the course.

9.2 The course

The uses, abuses and limitations of statistical information in a variety of activities, such as polling, public health, law, marketing, government policy and the environment, are examined via media reports such as the ones quoted above. The course is designed to prepare everyone, regardless of statistical background, to become critical consumers of statistical information. The course is designed to raise the statistical literacy level of aspiring journalists, politicians, sociologists,

Table 9.1 Content of Lies, Damned Lies, and Statistics undergraduate course.

Topic	Content
Media Reports	Comprehending and evaluating reports and visual displays of quantitative information, critical questions, measurement issues, writing reports
Surveys and Polls	Populations, samples, sampling variation, bias, variation in estimates, confidence statements, margin of error issues, survey design, ethics
Experimentation	Observational studies, experiments, problems of different conclusions
Risk	Types of risk, odds ratios, interpreting small probabilities, statistical and practical significance, false positives, causation/association
Statistical Reasoning	Heuristics, variability, nature of randomness, expected values, coincidences, prosecutor's fallacy, conditional probability, Bayes' Theorem, eye witness evidence

lawyers, health personnel, business people and scientists. Content is delivered within five topics (see Table 9.1) and is concentrated primarily on conceptual understanding rather than calculation. The course is assessed using a combination of assignments (30%), tutorials (5%), mid-semester test (15%), and final examination (50%). Students enrolling in the course range from first year through to fourth year.

9.3 Statistical literacy assessment framework

The term statistical literacy has not gained a consensus about its meaning in the literature. Gal (2002) argues that statistical literacy applies to data consumers and describes people's ability to interpret and critically evaluate statistically based information from a wide range of sources, and to formulate and communicate a reasoned opinion on such information. Wallman (1993: 1) defines statistical literacy as 'the ability to understand and critically evaluate statistical results that permeate our daily lives – coupled with the ability to appreciate the contributions that statistical thinking can make in public and private, professional and personal decisions'. Schield (2004a: 1), however, focuses on 'critical thinking about statistics as evidence for inferences' and believes that a course in statistical literacy 'should give students the ability to evaluate the strength of statistics as evidence in arguments about causation'. Gal, Wallman and Schield, in their definitions, refer to the data consumer aspect of statistical literacy. Watson (1997) initially developed a view of statistical literacy that centred on media reports and focused on the data consumer. She described a hierarchy of three tiers of statistical literacy: basic understanding of probabilistic and statistical terminology; understanding of statistical language and concepts embedded in wider social discussion; and challenging claims in the media. Recently she widened her definition of statistical literacy to include knowledge and experience of how data are produced (Watson, 2006). Ben-Zvi and Garfield (2004a) frame statistical literacy in terms of skills, such as being able to organise data and construct displays that may be used to understand statistical information. Since a data consumer needs statistical knowledge to be statistically literate, these widened definitions that encompass knowledge gained from experiencing statistical investigations are not inappropriate. Gal's conception of statistical literacy, however, is the most relevant and useful for our type of course.

Many people have written and theorised about statistical literacy (e.g. Gal, 2002), but few have reported on their statistically literacy tertiary courses including surveys of their students (e.g. Isaacson, 2005) and even fewer have conducted research on school students' interpretation of media reports (e.g. Watson, 2006; Merriman, 2006). A question remains, however, as to whether the knowledge gained from a typical undergraduate introductory statistics course is sufficient for statistical literacy or is even a primary goal. According to Gal (2002) and Schield (2004b), if critically evaluating other people's statistically based reports is not explicitly taught then students will remain statistically illiterate. Students also need to understand the statistical concepts, reasoning and data based arguments that permeate their everyday life and appreciate how evidence-based thinking contributes to public and personal decisions. Awareness that many everyday events can be thought of from a statistical perspective, including the heuristics people use when reasoning, is also part of statistical literacy (Tversky and Kahneman, 1982; Gigerenzer *et al.*, 1999; Pfannkuch and Wild, 2004; Utts, 2004).

Gal (2002: 3) states that cognitively a critical evaluation of statistically based information is predicated on the joint activation of 'a knowledge component (comprised of five cognitive elements: literacy skills, statistical knowledge, mathematical knowledge, context knowledge, and critical questions) and a dispositional component (comprised of two elements: critical stance, and beliefs and attitudes)'. Our assessment framework for statistical literacy (Figure 9.1) uses Gal's (2002) knowledge and dispositional elements as a base. To Gal's knowledge elements we add 'heuristics and fallacies' but categorise this element and the 'worry questions' element as the trigger knowledge elements. The trigger knowledge elements require knowledge but they also scaffold habits of mind, the

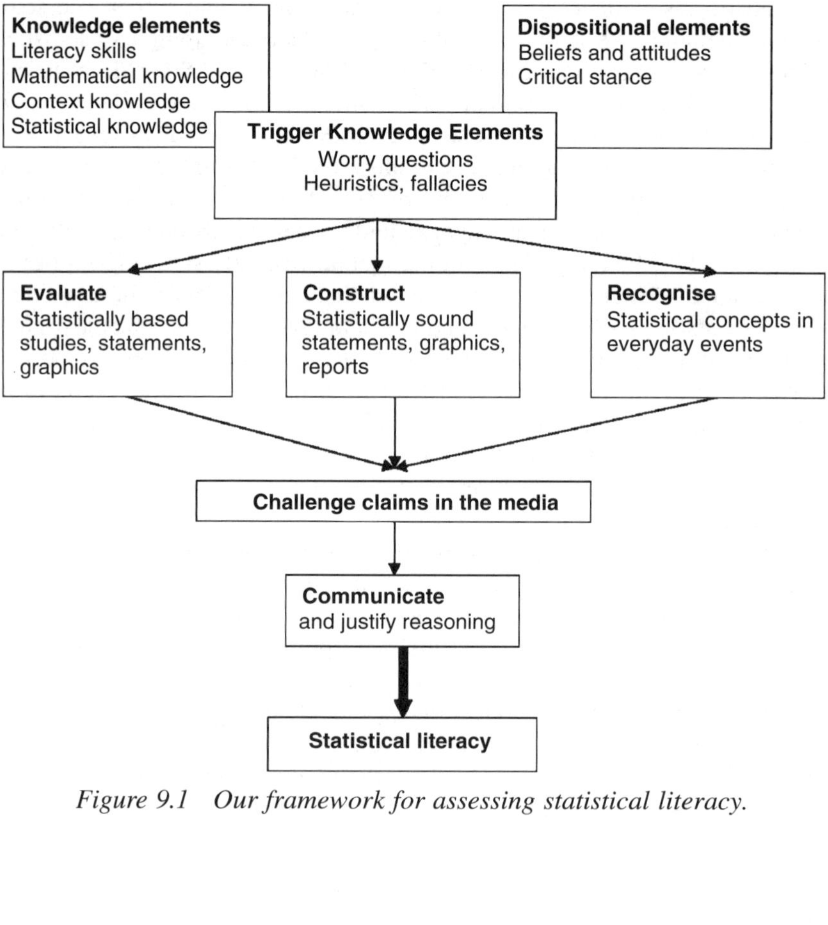

Figure 9.1 Our framework for assessing statistical literacy.

repeated use of which habituates and triggers students into adopting dispositions such as a critical stance and a willingness to challenge their beliefs. As well as the knowledge and dispositional components we assess five additional components. These are students' ability to evaluate, construct, recognise, challenge, and communicate, which are all interrelated but each needs to be specifically assessed as they require different cognitive skills. We will now elaborate on, in general, how we assess these five additional components.

From a knowledge and dispositional base, students critically evaluate statistically based information from a wide range of sources such as TV, newspapers, technical reports, and journal articles. When confronted with a newspaper article only, it is difficult for students to distinguish between critiquing the newspaper report and critiquing the actual study reported. Since a newspaper report typically provides little information about the underlying study, the worries about the study become hypothetical rather than based on actual fact. Therefore we actively encourage students, through our assessment, to evaluate the original source and then evaluate the media report upon it. Such an approach builds an understanding of how stories based on statistical data evolve: students learn that journalists may be slanting a story in a particular way or misconstruing the data; they also learn about the type of original source they might trust. Gelman and Nolan (2002: 77) also encourage their students to find the original source, as they 'found that the students are better able to evaluate the merits of the study and the quality of the reporting when given the original reports, even though the reports are often quite technical'. Critical evaluation involves analysing a report in detail and then synthesising that analysis into a judgement on the report. The judgement requires students to select the key factors from the analysis that will be used as evidence or justification for their opinion on the study.

The ability to synthesise statistical information from an original source into a summary report also needs to be specifically assessed. Working from their knowledge and dispositional base, the students construct statistically sound statements and graphics as well as writing a newspaper report based on an original source. We concur with Murray and Gal (2002), that this ability is different from critically evaluating a report as the students must create a report rather than respond to a report. Such a skill is particularly useful to university students as they may write reports in their future careers. Many of these reports will involve data-based evidence. Synthesis complements analysis and so students will learn more about critically evaluating reports if they are required to write them.

Using examples of everyday events, students are assessed on their ability not only to recognise concepts such as regression to the mean (Tversky and Kahneman, 1982) and Poisson clumping (Rosenthal, 2006) in real contexts, but also to recognise events where statistical concepts are not used but should be used such as eyewitness line-ups (Wells *et al.*, 2006). Everyday event knowledge also includes awareness of heuristics people use for reasoning about a multitude of everyday life events from a statistical perspective. The challenging of claims in the media is assessed in three ways: (1) as a consequence of students evaluating an original source and reflecting on the subsequent media article; (2) by students constructing a media report from an original source and then comparing their article to a media article on the same study; (3) by challenging claims directly reported in the media. If students challenge claims in the media, then they need to communicate and justify a reasoned opinion based on evidence. That is, after students have evaluated, constructed, recognised, or challenged claims, we assess their ability to give a reasoned opinion that is evidence-based, not based on their personal experience and pre-existing opinions. Thus, our assessment framework aims to assist students to become both critical consumers and communicators of statistically-based information whether from a media report, original source, or an everyday life event.

9.4 The assessment

Across the assessment tasks we ask students to evaluate media articles, journal articles and technical reports on opinion polls, sample surveys, experiments and observational studies. Since the course is relatively new we are still in the process of developing assessment tasks. For the type of statistical literacy we are attempting to foster, there are currently no trustworthy external measurement instruments to ascertain whether we are improving students' statistical literacy. Our assessment tasks may evolve into a valid measurement instrument. Hence, we can only describe and analyse the assessment tasks with reference to our model of assessment (Figure 9.1) and the statistical literacy literature for substantiation of the abilities that we may be assessing.

9.4.1 The first assignment

Assignment One requires students to critically evaluate a statistically based opinion poll or sample survey study. About three journal articles or technical reports are presented, together with related media reports, from which they select one. They are expected to evaluate and give a judgement of the study, and critique the related media report.

In discussing the abilities we believe we are assessing in this assignment, we use, as an example, a newspaper article entitled 'Doctors' white coats and facial piercings turn-off for patients' (Figure 9.2), and its associated journal article (Lill and Wilkinson, 2005). Briefly, the aim of the journal article was to investigate patients' preferred dress style of their doctors. The conclusion was that patients tend to prefer their doctors to wear semiformal attire and although most are comfortable with conservative items, many less conservative items are also acceptable.

Assessing ability to evaluate When students evaluate or analyse their choice of study they are encouraged to use a set of 'worry questions' or critical questions to prompt their thinking. These 'worry questions' are framed in Utts' (2004) seven-step template of seven critical components for evaluating news articles

Doctors' white coats and facial piercings turn-off for patients
NZ Herald, 21.01.2006

Patients like their doctors best when dressed in semi-formal clothes, but they are going off white coats, and facial piercings make them positively edgy, a study has found.

White lab coats and a formal suit and tie once held sway among doctors, but the Christchurch Hospital study reveals the new trends.

Visits to hospitals now show white coats have become a rarity, ties are far from universal and medical students sometimes push the boundaries of casual. At Auckland's Starship children's hospital, even the nurses were long ago freed of uniforms in favour of tidy casual dress.

The Christchurch study, published in the British Medical Journal, says white coats may be a source of cross infection between patients.

The study of 451 patients put long trousers, long sleeves and closed shoes at the top of the preference list for their male doctors, while dyed hair and facial piercing made them feel uncomfortable.

The patients felt most comfortable with their women doctors in long sleeves, with hair tied back and a long dress, and least comfortable if they had facial piercings or short tops.

Dr David Galler, a Middlemore Hospital intensive care specialist and adviser to the Director-General of Health and Minister of Health, is one doctor who takes unconventional dress very seriously.

"You've got to be able to carry it if you're going to be out of uniform," said Dr Galler, whose usual attire includes shorts and a bright shirt.

He believes doctors ought to normalise medicine for patients, rather than "hiding" behind a white coat and tie. He had never received complaints about his style and he thought patients liked it.

"What patients want is people who are honest, clean, respectful and knowledgeable," he said.

The Christchurch researchers, medical student Marianne Lill and associate professor Tim Wilkinson, found that even more popular with patients than doctors in semi-formal dress were doctors in the same style with the addition of a smile.

Figure 9.2 Example of newspaper article for Assignment One.

and studies. In response to the critical component on the individuals or objects studied and how they were selected, a student raised the following concerns:

> The outpatients were approached in a waiting room of various medical and surgical specialties. This would have excluded those people that were just about to be seen, or who arrived late for their appointments . . . The use of a convenience sample may have biased the results . . . The selection process of asking the outpatients in the waiting room means that the ones surveyed may have been the ones that arrived early for their appointment . . . they may be more likely to be the elderly.

Thus, the student recognised that a convenience sample was problematic. She was also worried about the mean age (55.9 years) of the sample and gave her own ideas on how such a situation could arise.

In the survey, six modes of attire were used in photographs to determine people's preferences: white coat, formal, semiformal, semiformal with smile, casual, and jeans. The top-ranked photo was semiformal with smile, followed by semiformal. The following paragraph illustrates a student's thoughts on the critical component about the nature of the measurements made and questions asked in the study. In particular, she suggests that a mannequin with outfits would be more appropriate, as the focus of the survey was on attire not the doctor:

> It was useful to provide a picture as opposed to a video clip to present attire, as in the video clip a lot more other factors come into play, such as doctors' mannerisms, environment, etc. In this regard, the measurements do really address the problem. It may have been more appropriate, however, to show just a mannequin with the outfits on, as the appearance of the male and female models used in the pictures may have influenced results.

Having read the study, the students raised many confounding variables it did not address, such as ethnicity and inconsistent use of colour of clothing in the photographs of doctors, the time of year the study was conducted, and a possible non-response bias. There is an expectation by some students that the 'answers' to the worry questions will be found in the study. But other students generate some really interesting insights on issues such as confounding variables and potential flaws in the design of the study. For example the study found that older people (>65 years) preferred the white coat. Hence a student thought of alternative explanations for Step 6 of the 'worry questions' template for this finding:

> Opinion of doctor attire may differ for more than just the age of a patient. Older people may be at hospital for different reasons, such as the severity of their condition, so expect more professional attire amongst their doctors. So it may not be that older people are more likely to support conservative dress but rather people who are at the hospital for a serious reason might like conservative dress, and there may be a correlation between older people and severity of cause. It is this sort of extraneous factor that may be an explanation for the results, rather than just on the basis of age.

To evaluate a study students need to use their contextual knowledge to think of, for example, why a sampling method might be biased, why the measures used may not address the problem, and plausible alternative explanations for the findings. In this way the context knowledge element is assessed.

Assessing ability to communicate After students have completed the analysis they are assessed on their ability to provide a reasoned opinion on or judgement of the study. Some students find it difficult to form a judgement and justify it by selecting and summarising key issues from their analysis of the study. This reasoned opinion necessitates formulating a synthesis of the information they have gleaned from their analysis using the 'worry questions' template. Some justify their judgement by resorting to their own beliefs or experiences rather than giving justifications based on the study. Others introduce some interesting and thought-provoking issues based on their own real-world knowledge, and from thinking laterally. Some put in additional effort to back up their views:

> They [the authors of the study] state that 'patients prefer doctors to dress in a semiformal style, but when accompanied by a smiling face it is even better, suggesting a friendly manner may be more important than sartorial style'. The study focused on the appropriate dress attire, not a doctor smiling. There was only one photo of a smiling doctor in 'semiformal' attire, so this conclusion cannot be reached unless all of the photos were accompanied by a corresponding smiling one. Also, as stated earlier, the corresponding photo should be exactly the same outfit but with a smile, as this may change results.

From our experience, we would agree with Murray and Gal (2002) that the analysis and synthesis of statistical information are two interrelated components of statistical literacy but both need to be specifically assessed.

Assessing ability to challenge claims in the media After critically evaluating the study, the students then consider the newspaper article on the study and determine whether the claims are justified. For example, the article on 'Doctors' white coats a turn-off for patients' (Figure 9.2) generalised the study to all New Zealanders, despite the fact that the study was conducted in only one city, and the claim was not true for older age groups. The former point was noted by a student:

> I would emphasise that the results from this hospital are not reflective of the general New Zealand patient opinion. Further I would make note that the study included a high level of older patients. This may be disproportionate, but further investigation needs to be done to find whether this is the case as it may in fact be reflective of average patient population as old people are more likely to go to hospital.

Hence by first evaluating the original study, we believe that students have a more substantive base from which to challenge claims in the media.

9.4.2 The second assignment

The goal of the second assignment is to have the students construct a statistically sound report. Students are asked to find a newspaper article that reports a study that is of some personal interest. They are asked to track down the original source and summarise the study and its findings. They then write their own newspaper article, together with a graph or table, and compose a letter to the editor of the newspaper stating whether the results of the research were meaningful. Finally, they state how and why their newspaper article differs from the original. Since students choose their own article, they tend to select a study that is within their level of comprehensibility, thereby allowing them to work at their potential level. We now describe the abilities that are assessed in the assignment.

Assessing ability to construct statistically sound statements The assessment schedule for the newspaper article, which is given to students in advance, assesses their ability to include the main facets of the study for comprehensibility, to outline potential limitations of the study, and to create a statistically sound graphic and statements. The types of graphics and statements the students are required to use are not found in typical newspaper articles. For example, the graphics need to indicate variability in estimates by using confidence intervals and in their text confidence interval statements are expected to prevent over-interpretation of numerical information and to appreciate that sampling error is associated with almost all statistical summaries. Writing appropriate non-causal statements, defining the population on which inferences are made, differentiating between sample and populations statements, clearly defining the measures used, and so forth, are checked for correctness in the assessment of their newspaper article.

Figure 9.3 gives an example of a typical newspaper article written by a student in the course. To illustrate how students write or do not write statistically sound statements and to highlight some of the main points we assess, the following can be noted about this article. Points 1 to 12 are linked to the superscripts within the newspaper example.

1. The title is a causal statement. A more appropriate title would be: Bad bosses may bring blood to boil.
2. Statement is not a valid conclusion, as it is not drawn about the population. The student has personalised the main finding of the study. A better statement would be: A British study has found that having a bad manager may increase the blood pressure of British female healthcare assistants.
3. The previous three paragraphs give sufficient information about the rationale for conducting the study and the study method.
4. The standard deviation is included in this sample statement as well as stating clearly the groups being compared.
5. A confidence interval statement, equivalent to a margin of error quoted for polls in newspapers, is made.
6. Students use the device of 'consulting an expert', a fictitious person, such as the Dr Serj Yin in this student article, to raise more issues about the study.
7. The strength of the evidence of the findings is conveyed.
8. Many students refer to sample size as a limitation, a facet we discourage. A better statement would be: It is unfortunate that about half of the participants withdrew from the study as these people were reported to have different personality characteristics compared to those who completed the trial.

Bad bosses bring blood to boil[1]

By Student

A British study has found that having a bad manager may increase your blood pressure.[2]

The experiment carried out by researchers at the Buckinghamshire Chilterns University College, amidst concerns about the huge number of deaths resulting from coronary heart disease. High blood pressure, along with high cholesterol and cigarette smoking, are the major risk factors for coronary heart disease.

28 female healthcare assistants who rated their supervisors in a questionnaire took part in the study. 13 of them alternated between a supervisor who was liked and one who was not, while 15 others formed a control group, who worked under either one supervisor, or two similarly perceived supervisors.

Reading were then taken every 30 minutes from blood pressure monitors worn by the women over a day under each supervisor.[3]

The study was published in the *Journal of Occupational and Environmental Medicine*. It found that the participants showed a mean increase of 15mmHg on average in systolic blood pressure (SBP) with a standard deviation (sd) of 11.9mmHg, and 7mmHg (sd=5.4) in diastolic blood pressure (DBP) when working under the disliked boss compared to a favourable one.[4] By contrast, the control group showed only a 3mm difference in SBP and 1mm difference in DBP between the two days.

In particular, it is estimated the SBP increases on average somewhere between 6.5 and 22.9mmHg for female healthcare assistants when working under a bad boss compared to working with a good boss. This statement is made with 95% confidence.[5]

Higher blood pressure has been proven to lead to greater risk of coronary heart disease.

However, Dr Serg Yin, a cardiologist at Heartville Hospital emphasises that someone experiencing a slight increase in blood pressure without any other risk factors such as smoking or obesity is still considered at low risk of suffering from coronary heart disease.[6]

What's more, he suggests, that a bad boss is unlikely to affect everyone in the same way – and a wide range of results, from a difference of 6 to 37mmHg for SBP, for the different participants in the study vindicate his view.[7]

He also remarked that "it is unfortunate that such a small sample was used."[8]

Dr Yin is currently planning a survey of over 300 heart disease patients at Heartville Hospital to find out several aspects of their lifestyle – including their work and attitude towards their supervisors, which may prove the findings of the British study more conclusively.[9]

Those concerned about excessive stress in their lives are advised to take part in yoga or meditation to help lower their blood pressure.[10]

11

Systolic BP: the force that blood exerts on the artery walls as the heart contracts to pump out blood. **Diastolic BP**: the force as the heart relaxes to allow blood to flow into the heart[12].

Figure 9.3 Article written by student in course.

9. Students are required to draw on their own knowledge and the journal article to think of alternative explanations or other factors that should be considered for the study findings.

10. Recommendations are given based on the findings.

11. Graphics need to show variability in the estimates such as drawing standard error bars on bar graphs, which is typical of many journal articles. We would prefer students to draw confidence intervals only but they are not penalised

for bar graphs, as the graphic must be easy for someone else to understand. Also, students need to create their own graphic or table from the journal article as done here, not copy one directly. Note the error bars added by the student have been incorrectly calculated in this example.

12. The measures SBP and DBP are defined.

Assessing ability to challenge claims in the media and communicate a reasoned opinion After writing their newspaper article, the students analyse how and why it differs from the journalist's article. This task assesses their ability both to critically evaluate the journalist's article and to state the underlying statistical principles that make the statements in their newspaper article sounder. For example, the student who wrote the article in Figure 9.3 mentioned that the journalist's report stated: 'when they worked under someone considered fair, everyone relaxed and their blood pressure dropped'. When the student looked at the statistics in the original report, she discovered that this statement was inaccurate and therefore felt that her article had clarified that 'there were wide variations across different individuals in the experimental group – indicating that this is perhaps not something that affects everyone'. From our experience, this task identifies students who have a better understanding about what makes statements statistically correct, as they must justify why statements are statistically sound or not sound.

9.4.3 The mid-semester test

In addition to assessing the same abilities in the test as described for the first assignment, we also assess knowledge that we believe is essential for being able to understand media reports. If students do not have some basic knowledge about representative samples, confidence intervals, significance, experimental and observational study designs, risk, and so forth, they will find it difficult to comprehend a study. The following two examples illustrate the type of knowledge and abilities we are assessing.

Example 1: Assessing ability to ask further questions about a claim Quotes are given to students to query such as: 'Since 2002 the unemployment rate has dropped by 20%'. By excluding data collection issues we expect students to query the starting value, how unemployment has been defined and whether the criteria for unemployment has changed in that period. Part of statistical literacy is to engender a healthy scepticism about claims made and to develop a feeling for what should be measured and compared to give a fair representation of a situation and to encourage curiosity. Prompting an array of questions is part of developing an ability to take a critical stance.

Example 2: Assessing statistical knowledge and ability to comprehend an article The students read a newspaper article reporting on some aspect of risk and answer questions. Implicit in the assessment is that the students require a certain level of literacy skills to make sense of the article together with their statistical knowledge. For example, newspaper articles can assess their knowledge on:

- Experimental design.
- Explanatory and response variables.
- Relative risks which should be accompanied by baseline risks.
- Possible confounding variables where they need to draw on their real-world knowledge.

Contextual knowledge, as well as ability to understand statistical terminology, is also assessed. By using a newspaper article, we are assessing whether students can cope with the types of statistical information that everyone is exposed to all of the time. These include ideas of randomised experiments and observational studies, and when we can and cannot conclude that a factor is the cause of an effect. We want to assess their ability to see weaknesses in the way that studies are performed so that they can discount sources of information which are obviously biased.

9.4.4 The final examination

Approximately half of the final examination paper is constructed around a newspaper article and its associated journal article both of which are given to students. The remainder of the examination consists of both short- and long-answer questions on the statistical reasoning topic. The following examples illustrate the types of abilities and knowledge we are assessing. For Example 3, the newspaper article was 'Under 30-s "highest risk group" for DVT' (*NZ Herald*, 26/9/2007) and the journal article was 'The Absolute Risk of Venous Thrombosis after Air Travel: A Cohort Study of 8,755 Employees of International Organisations' (Kulpers *et al.*, 2007). Example 3 uses information from Figure 9.4.

Example 3a: Assessing statistical knowledge

a. The journal article states that the researchers conducted a cohort study among employees of large international companies and organisations, who were followed between 1 January 2000 and 31 December 2005. What is the name of this type of study and what does it involve?

b. Describe any problems that might occur in classifying women into the two categories 'Oral Contraceptive (OC) No' and 'OC Yes' in Table 2 (Figure 9.4).

c. A potential problem with observational studies is extending results inappropriately. Explain why extending the results inappropriately might be a problem in this observational study.

Table 2. Incidence Rates, Absolute Risks, and Incidence Rate Ratios within 8 Weeks of Long-Haul Flights, for the Whole Study Population and Stratified by Sex, Age, Oral Contraceptive Use, Height, and BMI

Category	Air Travel[a]	Cases	Person-Years	IR/1,000 PY (95% CI)	IRR (95% CI)[b]	Flights	Risk/Flight[c]	Case/Number of Flights[d]
All (8,755)	No	29	27,772	1.0 (0.7–1.5)	Reference			
	Yes	22	6,872	3.2 (2.0–4.7)	3.2 (1.8–5.6)	102,429	21.5	1/4,656
Men (4,915)	No	12	14,728	0.8 (0.4–1.4)	Reference			
	Yes	13	4,810	2.7 (1.4–4.4)	2.7 (1.2–6.0)	76,461	17.0	1/5,882
Women (3,819)	No	17	12,968	1.3 (0.8–2.0)	Reference			
	Yes	9	2,050	4.4 (2.0–7.8)	3.3 (1.5–7.5)	25,780	34.9	1/2,864
<30 y (1,392)	No	3	4,132	0.7 (0.1–1.8)	Reference			
	Yes	3	616	4.9 (0.9–12.1)	7.7 (1.6–38.4)	8,014	37.4	1/2,671
30–50 y (6,017)	No	17	19,576	0.9 (0.5–1.3)	Reference			
	Yes	15	4,879	3.1 (1.7–4.9)	3.7 (1.8–7.5)	73,624	20.4	1/4,908
>50 y (1,345)	No	9	4,063	2.2 (1.0–3.9)	Reference			
	Yes	4	1,376	2.9 (0.7–6.5)	1.4 (0.4–4.6)	20,791	19.2	1/5,198
OC[e] No	No	9	10,193	1.0 (0.5–1.8)	Reference			
	Yes	3	1,533	2.3 (0.4–5.6)	2.2 (0.6–8.1)	18,085	20.3	1/4,938
OC[e] Yes	No	5	2,367	1.9 (0.6–3.9)	Reference			
	Yes	3	436	6.6 (1.2–16.4)	3.6 (0.8–14.9)	7,695	55.3	1/1,808
<165 cm	No	5	7,284	0.7 (0.2–1.5)	Reference			
	Yes	7	1,108	6.3 (2.4–12.0)	9.8 (3.1–30.9)	14,250	49.1	1/2,036
165–185 cm	No	21	16,759	1.3 (0.8–1.9)	Reference			
	Yes	11	4,602	2.4 (1.2–4.0)	1.9 (0.9–3.9)	69,095	15.9	1/6,281
>185 cm	No	3	3,493	0.9 (0.2–2.1)	Reference			
	Yes	4	1,115	3.6 (0.9–8.1)	3.7 (0.8–16.9)	18,242	21.9	1/4,561
BMI <25	No	16	14,919	1.1 (0.6–1.7)	Reference			
	Yes	7	3,617	1.9 (0.7–3.7)	1.9 (0.8–4.7)	51,958	13.5	1/7,423
BMI >25	No	13	12,546	1.0 (0.5–1.7)	Reference			
	Yes	15	3,198	4.7 (2.6–7.4)	4.9 (2.3–10.6)	49,509	30.3	1/3,301

[a]No,no exposure to air travel within 8 wk; yes, exposure to a flight of at least 4 h.
[b]IRR adjusted for age and sex.
[c]Risk per flight: risk per 100,000 flights.
[d]Number of flights needed to cause one case.
[e]Oral contraceptive use amongst women <50 y.
doi:10.1371/Journal.pmed.0040290.t002

Figure 9.4 Data from journal article, The absolute risk of venous thrombosis after air travel (Kulpers et al., 2007: 1511).

These items are designed to prompt students to think about how statements in the newspaper article may be misleading and to give a reason based on statistical knowledge. We believe that we are assessing their statistical knowledge, how statistical argumentation is constructed, and are indirectly developing the disposition to take a critical stance.

Example 3b: Assessing mathematical knowledge

> Using Figure 9.4, explain, showing your working, how the IRR of 3.3 was obtained for women. Describe in words what the IRR/1,000 PY numbers 1.3 and 4.4 and the IRR of 3.3 represent for women.

Basic mathematical knowledge is needed to be statistically literate. Anecdotal evidence from tutorials alerted us to the fact that students did not understand an increased risk such as 100% and had misconceptions about rates by thinking that the base rate was the number of people sampled. Therefore this item tests whether students understand how the figures were obtained.

Example 3c: Assessing construction of statistically sound statements

> The newspaper article reports that 'long haul passengers are three times more at risk of DVT than those who do not fly'. Using Table 2

(Figure 9.4), write a statement that reports on the confidence interval for this relative risk. Write your statement for a newspaper audience.

This item assesses students' ability to construct statistically sound statements and that when they write, for example, a relative risk statement they should include the baseline risk and a confidence interval.

Example 3d: Assessing ability to communicate a reasoned opinion on a study

Write a serious letter (about 200 words) to the *NZ Herald* editor in response to the newspaper article 'Under 30-s 'highest risk group' for DVT'. State you have read the original source. State one misleading claim in the newspaper article and provide a corrected version. Outline any potential limitations of the study. Include in your letter whether you are willing to believe the findings or whether you need more convincing. Justify your judgement and any other statements that you make. End your letter with your recommendations and advice to potential air travellers in New Zealand.

We ask students to write a letter to the editor, which again assesses their ability to construct a reasoned opinion on statistical information. Scaffolding in the examination questions assists students in the production of a letter, since they have been asked to think of limitations, confounding variables and worries about the study. But in the letter they must synthesise the information and rationalise whether the study should be taken notice of by potential air travellers. With the requirements for a recommendation and willingness to believe the findings, the beliefs and attitudes of the students tend to be exposed and hence the ability to supply evidence-based rationales is also assessed.

Example 4: Assessing ability to reason statistically about everyday events, identify fallacies and construct correct statements

a. An investment company made a return of 13% last year on its clients' investments, whereas the average return for all investment companies was 7%. The company attributed its success in being one of the best companies in Auckland last year to its wise investment practices. Explain, using the principle of regression to the mean, why the company's reason for success may be ill-founded. Draw a graph as part of your explanation.

b. In May 2005, New Zealand school children were being vaccinated against meningococcal B. At the same time there was a flu outbreak that kept many pupils at home. Some parents believed that their children's flu was caused by the vaccination. Consequently other parents refused to have their children vaccinated. (*NZ Herald*, 31/5/2005).

Explain to the parents who refused to have their children vaccinated why such an occurrence is not so amazing.

c. A newspaper headline read 'Black weekend on roads claims 10 lives' (*NZ Herald*, 23/8/2004). Explain why some people think such a cluster of road deaths is unusual. Explain to someone who has never studied statistics how such clusters can occur based on random variation alone. Illustrate your explanation with an example.

Example 4 tests students' awareness of heuristics – common misconceptions and fallacies in statistical thinking and reasoning that occur in everyday life.

We have provided exemplars of how we assess students' statistical literacy according to the components and their constituent elements listed in our assessment framework (Figure 9.1). Our approach to improving statistical literacy is based on the belief that:

- Using statistically based reports drawn from the New Zealand media will act as a catalyst for learning and improving statistical literacy skills.
- Critiquing the original sources will assist students to evaluate media articles and provide them with a better ability to communicate a reasoned opinion.
- Writing a report from an original source will help students develop the skills to construct statistically sound statements, recognise when statements are not sound and enhance their ability to challenge claims.
- Acquiring basic knowledge about surveys, experiments and observational studies, including specific knowledge (for example, about how to calculate and interpret relative risk and how to interpret confidence intervals from original sources) is necessary for students' comprehension.

Exposing students to many different and diverse contexts where statistics is used in research will help them recognise and build understanding of statistical concepts and 'transfer thinking learned in one applied context to other contexts ... abstraction-as-process' Cobb (2007: 338). Statistical literacy means being able to use statistical thinking as a way of viewing the world. Our assessments are designed to equip students with the ability to be critical consumers of statistically-based information, whether from media reports, original sources, or everyday life events, and although the dispositional component is not explicitly assessed, we believe that these assessments prompt the ability to take a critical stance and challenge students' beliefs. Until valid means are devised for measuring the type of statistical literacy we aim to foster, we cannot ascertain the degree to which our students have improved their statistical literacy as a result of participating in the course.

9.5 Do our students think they improved their statistical literacy?

The participants in this pilot study were four students, whose final course grades ranged from a low C to a high A. Three, typical of the student profile on the course, were not majoring in statistics and were drawn by general interest, intrigued by its title: 'Lies, Damned Lies, and Statistics'. The two students who had taken the course in 2005 had undertaken an introductory course in statistics; the student who had taken the course in 2004 had no additional statistical background; and the student who had taken the course in 2006 was majoring in statistics, and had taken the paper because she thought it sounded interesting and would be applicable in real-world situations.

The participants were interviewed by the first author, who subsequently analysed video footage of the interviews, each of which lasted 30–40 minutes. To stimulate their thinking they were asked about their reactions to two newspaper articles presented to them at the interview – one on a political poll, the other on a report about risk. All of the participants responded appropriately to the newspaper articles, which demonstrated that they retained knowledge from the course. This knowledge included taking into account the margin of error and the impact of comparing subgroups on the margin of error when reading media reports about political polls, and issues such as missing baseline risk when evaluating articles on risk. They were also asked about their use of statistical thinking and reasoning in their everyday lives.

Three main themes emerged from the interviews: the usefulness of the 'worry questions'; the importance of the original source when reading a newspaper article; and the impact of the course on the way they viewed the world. All participants told us that the 'worry questions' provided a valuable template, which helped them to challenge the claims made in media articles. This was noticeable when they were reading the given newspaper article, which was based on an observational study on caffeine consumption and risk of miscarriage. They questioned causal claims and identified potential confounding variables where links between the explanatory and response variables were made, queried measurements that were taken, and challenged the reliability of statements such as 'Women . . . never changed their caffeine consumption during pregnancy', which most found hard to believe. One participant admitted that, before taking the course, while she had been sceptical of many newspaper articles, she 'would never have thought of looking for the original source . . . or have known what questions to ask, or where to go to look further'.

With regard to the newspaper report on a political poll, participants questioned the impact of the time at which the poll was taken, with the thought that political issues that were happening then might influence the poll percentages. They also queried the sizes of sub-groups, as this would affect the margin of error when making comparisons between the smaller political parties. Another issue raised was the way the samples were selected, one student recalling that, prior to the 2005 election in New Zealand, 'some market research companies incorrectly predicted the final results for the election'. She thought that the way 'they selected the participants wasn't as representative as it could have been. In fact, they ignored ethnicity'.

Three participants stated that they were reading newspapers quite differently from before taking the course. They had an awareness that headlines could be alarmist and might not correspond to the body of the article, or to the original source; they now sometimes sought the original source behind a report in order to find out additional information. While some participants had been sceptical of many media articles prior to taking the course, they felt that they now had the confidence and skill to access original sources, ask questions and search for answers, using their own background knowledge.

The participants also gave examples of how that the knowledge that they had gained from the course filtered through to their everyday lives:

- Serving on a jury, a participant had found that the statistical reasoning component of the course had enhanced her awareness of the issues surrounding court evidence, and that her assessment of the evidence was 'definitely influenced by the course'.
- As a result of the risk component of the course, a participant was able to explain to a smoker friend advised not to take the oral contraceptive pill, that she needed to ask: what her current risk of deep vein thrombosis was as a smoker; by how much it would increase when on the contraceptive pill; and by how much it would increase if she were to become pregnant. Another participant reported that sharing her statistical knowledge with co-workers often opened up searching discussions about newspaper articles and television programmes.
- Statistical thinking had been 'really useful' 'all through a postgraduate diploma' that had required a participant to write three research reports: 'in a statistically correct way', for which she 'got good marks'.

As a result of participating in our statistical literacy course, these students seemed to have developed a knowledge base that enables them to challenge claims, the ability to recognise statistical ideas embedded in many different contexts and the confidence to question and form an opinion on statistically-based studies. Although we cannot draw any definitive conclusions from such a small, non-representative sample, these interviews suggest that the course and assessment may be having the desired outcome for some students.

In response to our standard student evaluation surveys conducted at the end of each course, more than 80% of respondents stated that they learnt a lot in this course and that it had sharpened their analytical skills.

9.6 Conclusion

Assessment portrays to students what knowledge is important to learn, what skills are valued and the nature of the subject. Often there are mismatches between the desired curricula and the assessed curricula (Niss, 1993). The goals for our course are to:

- Instil in students the ability to 'think statistically'.
- Enable students to critically evaluate statistically based reports.
- Teach students to construct statistically sound reports.

We believe that our goals are closely aligned to our assessment and that by privileging media articles as the source of our assessment that we are portraying statistical literacy from an everyday data consumer perspective. Gal's (2002) cognitive components for statistically literacy underpin our assessment framework (Figure 9.1) and Watson's (1997) final tier in her statistical literacy hierarchy, challenging claims in the media, is incorporated into the framework. Hence, our framework is built on research findings. Further refinements to the framework could be made by considering the nature of statistical argument and incorporating it as a component of statistical literacy. A challenge we face is to develop a valid statistical literacy research instrument, which can determine not only the level of statistical literacy that students have achieved but also uncover gaps in their reasoning processes.

As a result of participating in our course, students may not be fully statistically literate but they will be aware of issues underpinning statistically based information in the media and in everyday life. Part of this awareness is prompted by our assessment method, which strives to meet the goals of our course.

10

An assessment strategy to promote judgement and understanding of statistics in medical applications

Rosie McNiece

10.1 Introduction

There are several stages to statistical analysis of data within medical investigations and the statistician undertaking such analyses must have a comprehensive understanding of the various steps involved. Typically, this might require involvement at the planning stage, in the data collection process, through to the statistical analysis of the data. A further important and often overlooked aspect of the statistician's role is to interpret the results of analyses and present the findings. The statistician must have the ability to express the results of statistical analysis clearly, within the context of the study at hand and in a way that can be understood by an audience with little or no statistical expertise.

This chapter gives details of a research-based assessment exercise which is aimed at deepening students' understanding of the processes involved in conducting analyses of medical data, and at developing the skill of communicating statistical information in a clear and comprehensive manner. The exercise has also proved to have the added benefits of promoting independent study and research skills and in reducing the opportunity for plagiarism between students.

Assessment Methods in Statistical Education Edited by Penelope Bidgood, Neville Hunt and Flavia Jolliffe

10.2 Background

We encounter statistics on an almost daily basis in many areas of our day-to-day lives, including business, politics, health and education. For example, the government targets, performance indicators and 'acceptable' standards that we hear, see and read about every day are usually based on statistics – more explicitly, statistical analysis of data. Such statistics often form the basis of change and reform to 'improve' many aspects of our lives. Hence, we would hope that statistics in the public domain are reliable, having been achieved through correct and appropriate statistical analysis.

However, there have been many instances of statistics being used wrongly, being misinterpreted or being misquoted – whether in error, or in order to portray something other than what the statistical evidence suggests. In some instances, the consequences of such misuses of statistics may do no more than create a false impression but, more worryingly, there have been several incidences where the effect of incorrect statistics has had a much more serious outcome. Many high profile examples of statistics gone wrong have occurred in the field of medicine where the consequences of poorly conducted statistical analysis has led to a serious and adverse outcome. One of the best documented of such incidences is the reported link between the Measles, Mumps and Rubella vaccine (MMR) and autism (Wakefield *et al.*, 1998), which was based on a very small study sample. Although the evidence seemed compelling at the time, a later review highlighted that the study design was fundamentally flawed and the majority of the original authors subsequently retracted their findings (Murch *et al.*, 2004). Another high profile example is the case of Sally Clarke, who was wrongly convicted of killing two of her children (Batt, 2004). The verdict was later overturned on appeal, the major argument of the appeal being that the calculation of the risk associated with double cot death within a family was incorrect (http://www.sallyclark.org.uk/RSS/html). The statistical 'mistake' was that the chance of double cot death, which was stated as 1 in 73 million in the original case, was incorrectly calculated. It was later suggested that the true risk of double cot death within a family was somewhere closer to 1 in 200 (http://news.bbc.co.uk/1/hi/health/4685511.stm). The severity of this error prompted the president of the Royal Statistical Society to write to the Lord Chancellor expressing the need for correct and informed statistical analysis in such situations (http://www.therss.org.uk/archive/evidence/sc_letter.html). There are many other incidences of misused and poor quality statistics such as these that have led to calls for improvement in the statistical 'quality' of medical research (Young, 2007; Altman, 1994).

The role of medical statisticians has multiple aspects. Not only must they be technically competent in identifying and undertaking correct analysis procedures but they should also be effective communicators. They need to be able to discuss and defend methodologies used, assess how much information resulting statistics can impart and identify any limitations of the analysis procedure. Furthermore they ought to be able to present their findings, particularly the statistics, within the context of the study in a way that is clear and comprehensible both to fellow statisticians and to those with little to no statistical expertise.

10.3 Teaching statistics in a medical context

In the teaching of statistics at undergraduate level, instruction in technical ability is vital but instruction in the wider role of a statistician is equally important. As teachers of statistics, we have an obligation to provide students with knowledge and understanding of theoretical and methodological concepts, practical applications of such concepts, and an appreciation of the many stages of conducting a statistical investigation, from problem formulation through to communication of results. As Gal and Garfield (1997a) summarise, there are several goals that should be achieved in teaching statistics and these encompass not only statistical methods but also the many aspects of the wider subject area.

This is particularly important in teaching applications of statistics in medical situations. In undertaking statistical analyses of medical data choosing and applying the appropriate methodology is essential but equally important is communication of the results (Strasak *et al.*, 2007). This requires the statistician to have a comprehensive understanding of the statistical and more general procedures involved in the data analysis process. Hence, in teaching students how to conduct statistical analyses of medical data, we should also teach them about the various stages in conducting a medical investigation and make them aware of all the steps involved in producing reliable, accurate and comprehensible information as and when required.

Our aim should be to produce competent statisticians who are prepared for the workplace. These persons should understand the different phases of a statistical inquiry and be confident in choosing correct methods to analyse data effectively. Crucially, they should also be able to present their findings within the context of the study at hand and in a manner that is comprehensible to those involved in the healthcare sector regardless of their level of statistical expertise.

10.4 'Introduction to Medical Statistics' – an undergraduate module

The module 'Introduction to Medical Statistics' discussed here is an undergraduate module and is a free choice option to students in the latter half of their degree programme. Students choosing the module will already have a thorough knowledge and understanding of the basic concepts of statistical inference and the testing procedures therein. They will have completed an introductory course in statistical modelling which covers both theoretical and practical application of the construction and validation of a general multiple linear regression model. They will also have been introduced to the concepts of interpretation and presentation of results and report-writing.

The module is designed to provide students with a basic introduction to the field of medical statistics. During the module, students are introduced to a variety of applications of statistics within the fields of healthcare and medicine, including demographic analysis and various types of epidemiological studies and clinical trials. On completion of the module, students should:

- be familiar with some of the basic statistical terminology and methodologies used in the field of medical research;
- be competent in identifying and analysing data from different types of epidemiological studies using appropriate statistical measures and models as necessary;
- be able to consider and discuss planning issues and ethical considerations involved in conducting epidemiological studies and basic clinical trials;
- be able to interpret and report clearly the results of any such analyses;
- have an insight into the practical problems involved in conducting studies in the medical field and the diversity of statistical techniques used.

The assessment tasks for this module have been designed to test and develop these skills.

10.5 The assessment strategy

Assessment of the module consists of in-course assessment and a traditional end-of-module examination. The in-course component is broken into two further parts, an in-class test and an individual assignment, both of which are completed before the end of the teaching schedule. The in-class test and formal examination test the ability to identify study types and the appropriate methods of analysing data, manipulation of data, calculations of appropriate statistical measures and basic interpretation of the results of such applications.

The second part of in-course assessment is an individual assignment. Students are asked to undertake some research to find published examples of two different types of epidemiological studies from those encountered during the course of the module and to then critically review various aspects of their chosen studies. They are asked to write a report to compare the different study types, to include discussion of why the different study designs are appropriate in each case and to discuss how the study results were achieved and presented. They are also asked to comment on any issues that might have arisen in the planning stages, including ethical considerations, and to discuss the limitations of the studies and thus by implication limitations of the study findings. A typical assignment question is as follows:

> You are required to find a real-life example of one cohort study and one case control study from published medical literature (you should browse books, journals, Internet and other relevant sources in your search). Then compare and contrast the two studies along the guidelines* outlined below, explaining why the chosen study types are appropriate for each investigation.
>
> (i) discuss the advantages and disadvantages of the methodologies in the context of the study.
>
> (ii) analyse the findings of the studies, including calculations where possible, comment on how the results of the study are presented and explain what the results mean;
>
> (iii) identify and discuss any limitations within the studies.
>
> (iv) suggest possible ways forward for further investigation into the link/causal relationship between the outcomes and exposures under investigation.
>
> You should also do some background research on the disease and exposures in your chosen study in order to set your discussions within the appropriate context. The discussion should be written in your own words – wholesale copying of an article will result in a zero score (You must include copies of the reports/articles with your submission).
>
> *Guidelines are not necessarily an exhaustive list. You should report on any aspects of the studies that you think relevant. Your report should not exceed 1500 words.

The assignment description is purposely left open to allow students some freedom and independence in producing their reports. Students are encouraged to think beyond the guidelines of the assignment, to consider aspects such as data collection and data analysis methods, and to add any information which they think relevant and that might further demonstrate their understanding of the use of epidemiological studies in practice. Students are usually allowed a period of four weeks to complete this individual assignment in order to give them sufficient time to conduct the research, contemplate the study protocol and submit a well planned and structured report.

The pedagogic aim of this assignment is to broaden students' understanding of the processes involved in conducting epidemiological studies. It is aimed at testing their appreciation of the wider implications of a medical investigation, which requires knowledge and understanding that cannot be assessed in a traditional examination. It is hoped that in completing this assignment students will think in depth about the planning, ethical considerations and limitations of a study, with the analysis of data being of secondary concern. The exercise should highlight and promote the importance of these skills in real life applications and has the added benefit of developing independent study and research skills.

10.6 Results of assessment exercise

In general, students have responded well to this assessment task, with some producing work of a very high standard. All have managed to complete the task of identifying real life examples of different types of epidemiological studies. The articles that have been submitted by students are drawn from a wide range of journal and online publications, including the more obvious sources such as the *British Medical Journal* (BMJ), *British Journal of General Practice, International Journal of Cancer* and *International Journal of Epidemiology*, through to disease-specific and less well-known publications, such as the *International Breastfeeding Journal, Annals of Rheumatic Disorders* and the *Malaria Journal*. The articles submitted also cover a variety of clinical topics where some form of statistical analysis is used. Examples include paediatric studies into childhood obesity and congenital birth defects, both acute and chronic diseases such as diabetes, asthma, lung cancer and heart surgery, and less clinical studies such as investigations of physiotherapy techniques in treating football injuries. Some of the more unusual submissions were:

- Mortality among workers at Municipal waste incinerators in Rome: a retrospective cohort study, 1997, American Journal of Industrial Medicine, 31:659–661
- Case-control study of indoor cooking smoke exposure and cataract in Nepal and India, 2005, International Journal of Epidemiology, 10.1093/ ije/dyi077
- A case-control study of farming and prostate cancer in African-American and Caucasian men, 2007, Occupational Environmental Medicine, 64: 155–160.

Most students have delivered at least fair attempts at the discussion aspect of the assignment and have provided good justifications as to why the study designs used are appropriate for the investigations being undertaken. Overall, the completed assignments show evidence that due thought has been given to the processes that underpin setting up a medical case-control or cohort study. Where possible, students have replicated the calculations used to produce statistics such as odds ratios from their chosen report. Sometimes this has required them to extract the relevant information from broader descriptions of the study where the actual tabulated figures were not expressed explicitly. Most have made some attempt to carry out background research relating to their chosen study; the extent of this varies greatly, with some students going into great detail and others barely skimming the surface. The aspect of the assignment relating to limitations of the study has not been addressed in great detail, which has been disappointing. On the whole, students do seem to have put in the effort required for this assignment, and so its aims are generally being achieved. In some cases, students have failed to meet the standard required and are encouraged to examine the work of peers who have produced better work, in order to be able to identify where they have not delivered.

Marking of the assignment has not, to date, been an onerous task for the lecturer. This is mainly due to the fact that the module numbers have been small, usually between 12 and 15 students, making the task of marking very manageable. The assignment contributes 25% of the total marks awarded for the module (compared to 15% for the in-class test and 60% for examination). The marking strategy is discussed with students at the outset. It is stressed that a basic pass can be achieved for correct examples of each study type along with a well presented discussion that addresses all four areas outlined in the problem statement. Higher marks will be awarded for evidence of thinking beyond the box, in other words where students have exceeded the guidelines set out. For example, high marks might be awarded where students produce a clear and detailed rationale for the study design, including discussion of why other study types would be inappropriate. Students are also reminded that there is a maximum word limit in order to prevent long and rambling discussions within submitted work.

A flexible approach to marking is used, based on a combination of set marks and general marks more akin to project marking. This flexibility allows the lecturer to get a feel for the submitted work and review marking of individual submissions as necessary by gaining an overall impression of standards. However, this is only achievable due to the small numbers involved and would become much more time-consuming if student numbers were to increase significantly.

Similarly, the small group size allows for monitoring of plagiarism between students and to some extent in a wider context; with small numbers involved, identical submissions are easy to spot. It is not unreasonable to expect that each student would choose a different study and submit an individual written report, although collaboration on research methods and resources is allowed and encouraged. Also, due to the prescriptive nature of the discussion aspect of the report, it is unlikely that this could be copied directly from other published sources. It is usually obvious when part of a discussion is not original to the student due to a noticeable change in the writing style and quality of prose, although it is possible that a particularly skilled student might be able to plagiarise. In addition, the lecturer keeps copies of all studies submitted with a view to eliminating passing on of work between successive academic cohorts. The course, and hence this assignment, has been running for three academic years and no incidences of plagiarism either between students or from other sources have been detected so far. Each student has managed to find their own study and all of these have been appropriate examples under the task set out.

10.7 Feedback from students

In general, student feedback about the assessment task has been positive. Several commented that they enjoyed the exercise and that it has helped them to appreciate the role of statistics in real-life applications, making what they have learned in lectures seem valuable and relevant to daily life. Some also commented that the exercise had helped increase their understanding of the issues involved in conducting epidemiological studies, as they had been required to think through the scenarios surrounding their chosen studies. The task is also aimed at developing research skills; in general students felt that this had been achieved and some even suggested that the exercise had impacted on their approach to other modules, as they have learned to appreciate the importance of communicating results. When students were asked about how well-prepared they felt for this task, there were mixed responses regarding their ability and competence to undertake the exercise. Some felt in hindsight that they had not fully appreciated what was expected of them.

10.8 Conclusion and further development of the assessment procedure

The rationale behind setting this exercise was to give students an understanding of the wider role of a statistician within any discipline, but particularly in the field of health and medicine. This is in response to my own, and more widely reported perception (Strasak *et al.*, 2007, O'Fallon, 2000) that the communication skills and thinking processes essential to the role of statistician are not always conveyed to students and are often lacking in graduates. One of the most important aspects of being a statistician is the ability to tell people what the numbers actually mean in real-life terms.

To date, this assessment task does seem to have had delivered what it was designed to do. In general the students have produced high quality reports covering a diverse variety of clinical conditions and have shown a good appreciation of the processes involved in conducting and analysing an epidemiological study. Most students have met or exceeded expectations and have clearly benefited from the exercise. In the few cases where they have failed in this task, it is usually due to lack of effort rather than lack of understanding of the task set. While it is difficult to evaluate deeper learning, the work submitted does provide evidence that students have, on the whole, developed an appreciation of the processes involved in undertaking a medical investigation over and above statistical calculations. Several of the students who have taken this course have subsequently chosen, and been successful in completing, a final-year project involving a statistical investigation of data.

The exercise has also been beneficial from a teaching point of view. It does seem to have minimised plagiarism as each student has to choose their own individual study and submit a report based on it. The exercise has also been useful as a resource-gathering procedure as the case studies submitted by students have been collected to create a portfolio of topical and interesting examples of epidemiological studies. A possible teaching development being considered for this module is to introduce more discursive sessions in the early stages, where students would be able to discuss different case studies within groups. The resources gathered from the assignments would provide useful material for these sessions. Such activities might help students better understand the expectation of the individual assignment. Further developments in the assessment of this module are also being considered. Possibilities include introducing a short oral presentation to the individual assignment, but this will be dependent on time restrictions and student numbers.

In summary, the individual assignment component of assessment for this module introduces students to the wider aspects of the role of a medical statistician. It aims to assess a deeper appreciation of the process of a statistical investigation than would be achieved through tests and examinations. It promotes thinking beyond producing figures based on calculations and increases students' understanding of the need for communication of results within the context of the study. It is hoped that the module will spark an interest in applications of medical statistics amongst students. At the very least, the skills developed in the assignment exercise should benefit anyone working in a statistician role across any discipline.

11

Assessing statistical literacy: Take CARE

Milo Schield

11.1 Statistical literacy: A new goal for statistical education

In 2006, statistical literacy was adopted as a goal by the ASA in endorsing the ASA Guidelines for Assessment and Instruction in Statistics Education (GAISE) Report (ASA, 2007). This goal is stated in the first sentence in the PreK-12 portion of the GAISE report:

> The ultimate goal: Statistical Literacy'. It is the first recommendation of the college section of the report: 'introductory courses in statistics should, as much as possible, strive to emphasise statistical literacy and develop statistical thinking' and it is in the ASA Strategic Plan (2008) for Education: 'Through leadership in all levels of statistical education, the ASA can help to build a statistically literate society . . . '

The PreK-12 section of the GAISE report underscores the importance of statistical literacy. 'Statistical literacy is essential in our personal lives as consumers, citizens, and professionals'. 'Statistical literacy is required for daily personal choices'. 'Statistical literacy involves a healthy dose of scepticism about "scientific findings"'. 'An investment in statistical literacy is an investment in our nation's economic future, as well as in the well-being of individuals'.

Assessment Methods in Statistical Education Edited by Penelope Bidgood, Neville Hunt and Flavia Jolliffe

11.2 Identifying statistical literacy

Broers (2006: 1) notes that the diverse attempts at defining statistical literacy (SL) 'usually take two different turns'. One stipulates 'what it is exactly, that a statistically literate citizen should know of statistics ... These discussions tend to be clear, although not very consistent'. The other focuses 'on additional requirements for becoming statistically literate; requirements other than SK [Statistical Knowledge] elements. Here, the discussion of what it means to be statistically literate becomes far less clear'.

The first approach begins by linking statistical literacy with 'for whom' (all liberally-educated adults) and 'for what' (to be good citizens and decision makers). The second links statistical literacy with cognitive skills that are selected based on expert insight.

Moore (1998: 1) follows the first approach when he focuses on the needs of individuals in different roles to distinguish statistical literacy from statistical competence. 'What is statistical literacy, what every educated person should know? What is statistical competence, roughly the content of a first course for those who must deal with data in their work?'

Gal (2002: 2) also follows the first approach when he defines statistical literacy by focusing on the needs of adults in modern societies. From their needs, he concluded that statistical literacy refers to two interrelated concepts, primarily (a) their ability to 'interpret and critically evaluate statistical information' which they may encounter in diverse contexts, and when relevant (b) their ability to 'discuss or communicate their reactions to such statistical information, such as their understanding of the meaning of the information, their opinions about the implications of this information, or their concerns regarding the acceptability of given conclusions'. Gal elaborates on this consumer-producer distinction, linking reading contexts with statistical literacy, and enquiry contexts with data producers. Although statistical literacy is considered a key educational goal, Gal (2003: 81) concludes that many instructors 'neither teach for statistical literacy nor assess it', that 'serious attention to statistical literacy issues (in terms of both skills and dispositions) cannot be accomplished within an introductory course focused on core statistical topics' and that 'separate courses focused on statistical literacy will have to be planned'. Utts (2003: 74) follows the first approach, in relating statistical literacy to what every educated citizen should understand about probability and statistics.

The college section of the GAISE report follows the first approach, linking statistical literacy with art appreciation and consumers of data. Some courses focus on teaching students to become 'statistically literate and wise consumers of data; this is somewhat similar to an art appreciation course'. Some courses focus on teaching students to become 'producers of statistical analyses; this is analogous to the studio art course'. The mixture of consumer and producer components 'will determine the importance of each recommendation'. This report suggested assessing statistical literacy by students 'interpreting or critiquing articles in the news and graphs in media'.

Ben-Zvi and Garfield (2004a) follow the second approach in trying to distinguish statistical literacy (SL) from statistical reasoning (SR) and statistical thinking (ST) in terms of the types of understanding or cognitive outcomes. Broers (2006: 4) questions the grounds of these distinctions and argues that in looking for learning goals, 'SL, SR and ST were postulated as constructs': constructs that could help us in searching for new directions in statistics education; constructs that could provide a better view of what we should be teaching and how best to teach it. 'Their existence is not dictated by empirical observations, but rather empirical observations are sought in order to justify their creation. It is the individual researcher who decides what will be included in a definition and what will be left out'.

Broers' psychological critique may be unjustified, but if a non-empirical classification is to avoid being arbitrary it must clearly distinguish the constructs and have strong support among subject-matter experts.

Despite the contrast between a user-needs approach and an expert-insight approach – between an empirical approach and an idealistic approach – they may give similar results.

A somewhat similar tension exists in the mathematics community in defining quantitative literacy. One approach focuses on finding applications of mathematical ideas in everyday life (Gilman, 2006), another on quantitative literacy as part of effective citizenship in a modern democracy (Madison and Steen, 2008; Steen, 2001).

In this chapter, the user-needs approach of Moore and Gal is combined with the effective citizenship approach of Steen and Madison to argue that statistical literacy should be empirically based on the statistical needs of educated adults in a modern society.

11.3 Statistical literacy: For whom

Statistical competence is the ability to produce, analyse and summarise detailed statistics in surveys and studies. Statistical competence is needed by data producers – students in quantitative majors that have a statistics requirement, such as business, psychology, sociology, economics, biology and nursing – and possibly majors that have a calculus requirement such as those in science, technology, engineering and mathematics (STEM).

Statistical literacy is the ability to read and interpret summary statistics in the everyday media: in graphs, tables, statements and essays. Statistical literacy is needed by data consumers – students in non-quantitative majors: majors with no quantitative requirement such as political science, history, English, primary education, communications, music, art and philosophy. About 40% of all US college students graduating in 2003 had non-quantitative majors (Schield, 2008b).

11.4 Statistical literacy: The goal

Once statistical literacy is generally defined, the next step is to identify what activities would serve as the best indicator of proficiency. Two closely related approaches are:

- Make a decision based on multiple statistics. Consider the Collegiate Learning Assessment (CLA) organised by the Council for Aid to Education (CAE). Students are asked to present a reasoned conclusion given either a simple performance task (analyse some summary data) or a 'Make an Argument' task (given some related factors). See www.cae.org/content/pro_collegiate.htm.
- Evaluate the statistics in the everyday media. Tables, graphs, headlines, news stories, press releases and research reports present statistical associations as evidence for causal connections (Schield, 2008a; Budgett and Pfannkuch, 2007; Madison, 2006; Lutsky, 2006; Hayden, 2004; Moreno, 2002; Snell, 1999; Watson, 1997).

Unfortunately, choosing between deciding and evaluating gives little guidance on what ideas, tools and skills should be assessed. Doing this requires input from subject-matter experts. Once these topics are identified, they can be ranked in two ways:

- Ranked by subject-matter experts. For such a ranking in statistics, see McKenzie (2004).
- Ranked by their prevalence in the everyday media.

11.5 Identifying relevant topics by subject-matter experts

Moore (1998: 1) thought that statistical literacy should involve two clusters of 'big ideas':

1. 'The omnipresence of variation, conclusions are uncertain, avoid inference from short-run irregularity, [and] avoid inference from coincidence'.
2. 'Beware the lurking variable, association is not causation, where did the data come from? [and] observation versus experiment'.

Best (2008) claims that the dominant influence on all statistics is human choice since all statistics are socially constructed. Not that reality is subjective, but that human beings decide what and how to count and measure, what to summarise and how to model, and what to compare and how to communicate.

Utts (2003) identifies seven topics that every educated citizen should understand about probability and statistics. These include distinguishing causation from association, experiment from observational study, statistical significance from practical importance, and no-effect from no-difference.

11.6 Take CARE

Schield (2008a) reviews the topics – the sources of influence on the value of a statistic – identified by subject matter experts. The goal is to classify these influences into a small number of categories that are exhaustive, exclusive and fundamental. Now, all too often there is something omitted so the categories are not exhaustive, there are borderline cases so the categories are not exclusive and identifying what is essential is certainly contextual so that in a different context the categories are not fundamental. Nevertheless consumers of statistics – people who are not working with statistics regularly – can benefit from focusing on a smaller number of categories, even if they are not logically pristine, provided they are fundamentally different.

All the factors that influence a statistic have been classified into four categories:

- **Context** The influence of factors taken into account (1) by counts, averages, ratios and comparisons of counts, averages and ratios; (2) by epidemiological models (cf., deaths attributable to obesity); (3) by regression models; and (4) by the study design (cf., controlled vs. uncontrolled; longitudinal vs. cross-sectional; experiment vs. observational study) or by selection (cf., in tables and graphs). The influence of related factors (confounders) that were not taken into account in the study and were not blocked by the study design.
- **Assembly** The influence of choices (1) in defining groups or measures, (2) in selecting the summary measure (e.g. mean vs. median), the type of comparison (e.g. simple difference versus times more), and the type of ratio (e.g. the confusion of the inverse or the prosecutor's fallacy), (3) in selecting the group in forming an average, the base in a comparison of numbers and the denominator in a ratio (e.g. rate or fraction) and (4) in selecting the graph, table or statistic in presenting statistical results and summaries.
- **Randomness** The influence of chance on averages and coincidences (e.g. hot hand, too unlikely to be due to chance and regression to the mean). The difference between statistical significance and practical significance in large samples or between 'no statistical effect' and 'no effect' in small samples. The influence of a confounder on statistical significance.
- **Error** (or Bias) The influence of any factor that generates a systematic difference between what is observed and the underlying reality: subject bias (people can lie), measurement bias (instruments can fail, questions may lead and researchers may manipulate) and sampling bias (the difference between the sampled and the target population influences the result).

Given the extensive influence of human choice on numbers, the W.M. Keck Statistical Literacy Project grouped these four sources of influence under the age-old admonition, 'Take CARE' where each of the four letters in 'CARE' signified a distinct source of influence on any statistic: Context, Assembly, Randomness and Error. If students were to remember to 'Take CARE' in analysing statistics, that would be a considerable achievement. The choice of 'Context' for the first category is based on the importance that context plays in the liberal arts and on the importance that statisticians place on context in distinguishing statistics from mathematics.

11.7 Relevant topics based on empirical evidence

But all this analysis is based on expert opinions. What about the empirical approach? Gal (2002: 19) noted that:

> What is basic knowledge . . . depends on the level of statistical literacy expected of citizens, [depends] on the functional demands of different contexts of action (e.g. work, reading a newspaper), and depends on the larger societal context of living . . . Unfortunately no comparative analysis has so far systematically mapped out the types and relative prevalences of statistical and probabilistic concepts and topics across the full range of statistically-related messages or situations that adults may encounter and have to manage in any particular society.

Here are the results of two empirical studies:

- An analysis (Schield and Schield, 2007) of 250 statistically-based news stories found that 75% were observational and thus vulnerable to confounder influence, 74% involved assembly (choices in definitions, comparisons or presentation) and over half involved the grammar of percent, percentage, rate or chance indicating the use of ratios.
- An analysis (Raymond and Schield, 2008) of 160 numerically-based stories found that 27% involve assembly in constructing categories or measures, and 62% present associations that imply causation. In terms of the four 'take CARE' categories, 42% of these stories involved statistics that were influenced by confounding, 17% by assembly, 11% by bias and 9% by random effects.

Although this is empirical data, these findings are very tentative. First, there is considerable latitude in identifying these influences in a news story. Second, these articles and studies are not very representative. They are based almost entirely on short health-related news stories. They do not include longer articles in magazines or the press releases or detailed reports of government statistical offices. Third, they involve the same author. Given the latitude in selecting the stories and in classifying these influence, having an independent analysis would be valuable.

Nevertheless, both these empirical studies uphold an important finding: that Context and Assembly are much more prevalent as statistical influences in the everyday news than are Randomness and Error/Bias. This finding is important in deciding what to assess and how to assess it.

This analysis of the definition, the nature, the goals and the content of statistical literacy is just theory – theory that is barren until it results in exercises that students can work and that are a useful basis for assessment. Without such exercises, statistical literacy will exist as a vague 'habit of mind' that cannot be readily assessed or taught. So let us turn to the assessment of statistical literacy.

11.8 Assessing statistical literacy: Introduction

Assessing statistical literacy can be done at four levels by having students:

1. Evaluate the use of statistics in a news story (Schield, 2008a).
2. Estimate a quantity or make a decision in an open-ended situation (Gilman, 2006).
3. Describe and compare statistics presented in graphs or tables.
4. Answer multiple-choice questions on specific aspects of statistical literacy.

The last two are presented below.

11.8.1 Describe and compare statistics presented in graphs or tables

To be statistically literate, one must be able to use ordinary English to describe and compare rates and percentages as presented in tables and graphs. This can be quite difficult for students – especially those for whom English is a second language. Most students do not see a difference between 'times as much as' and 'times more than'; most students do not see that 'the percentage of men who smoke' is the same as 'the percentage of smokers among men'; nor that 'the death rate of men' is the same as 'the rate of death among men'.

Assessing student classifications or writing can be done quickly using a grading template:

- *Part-whole classification.* When students are asked to identify the part and whole given a statistic in a description, graph or table, a three-, four- or five-mark scale is adequate (depending on the complexity of the ratio) with one mark deducted for each part-whole error. Consider this statement: 'Among women who were first married in 2003, the percentage who were ages 25–29 was 27%'. A student who classified 'first married' as part would lose one mark as would a student who classified 'ages 25–29' as whole.
- *Comparison of numbers.* When students are asked to write a statement comparing two counts, totals or averages, a four mark scale is adequate: one

for the proper comparison grammar (difference, ratio, or percent/times difference); one for the proper base indicator (as or than); one for the proper comparison amount; and one for the appropriate test and base. Suppose a student is asked to compare six and two as a ratio with the larger as the base, and they write, 'Six is three times more than two'. The student would lose three marks: one for using two as the base, when using the larger (six) was specified; one for using 'times more' grammar when a ratio (times as much) was specified; and one for using 'than' to introduce the basis of the comparison when 'as' is appropriate.

Statements that describe and compare ratios such as percentages and rates are more difficult to assess. Table 11.1, a table of percentages with 100% column totals, is used to illustrate the assessment process in such cases. The P and W letters signify Part and Whole respectively.

Table 11.1 Table of percentages (hypothetical data).

Students [W]	---------------SEX-[W]--------		
Ⓟ MAJOR	[W] MALE	[W] FEMALE	[W] ALL
Ⓟ Business	↓ (60%)	↓ (20%)	↓ 40%
Ⓟ Economics	↓ 10%	↓ 50%	↓ 30%
Ⓟ MIS	↓ (30%)	↓ 30%	↓ 30%
ALL	100%	100%	100%

- *Description of a ratio.* When students are asked to write a statement describing a ratio in a graph or table, a four mark scale is adequate – one mark for the ratio keyword and the rest for proper part-wholes. When asked to describe the 60% in the upper left corner using percentage grammar, a student wrote, '60% of business majors are males'. This reply would lose four marks: one for omitting 'college students'; one for describing 'males' as a part; one for describing 'business majors' as a whole; and one for using percent grammar instead of percentage grammar. A correct answer would be: 'Among college students, the percentage of males who are business majors is 60%'.

- *Comparison of ratios.* When students are asked to compare two ratios given in statements, a table or graph, a six-mark scale is adequate, with one mark for the keyword (percentage, rate, etc); one for part; one for wholes; one for test/base and base indicator; and two for compare (numeric and grammar). When asked to compare the circled 60% and 20% as a simple ratio using the smaller as the base using percentage grammar, a student wrote: 'Among college students, males are three times more prevalent than females among business majors'. This reply would lose five marks: one for describing 'business majors' as a whole; one for describing 'males' and 'females' as parts; one for using a 'times more' comparison when a simple ratio was requested; one for using 'than' to introduce the base in the comparison; and one for using likely/prevalent grammar when percentage grammar was requested. A correct answer would be: 'Among college students, the percentage of business majors is three times as big among males as among females'.

For more on the grammar for describing and comparing ratios, see Schield (2004a).

11.8.2 Answer multiple-choice questions on specific aspects of statistical literacy

The rest of this chapter deals with multiple-choice questions similar to those administered to students in non-quantitative majors at Augsburg College. These are taken from Moodle exercises involving over a thousand questions. While open-ended problems, essays and portfolios provide more comprehensive forms of assessment that may be needed to identify higher levels of statistical literacy, right-wrong exercises are useful in assessing basic levels of statistical literacy, for minimising instructor grading time and for handling large-enrolment courses. These questions are presented in four categories: context, assembly, randomness and error/bias.

11.9 Assessing statistical literacy: Context

To understand the influence of context on a statistic, a statistically literate person must understand the features and benefits of different types of study designs (observational vs. experimental, longitudinal vs. cross-sectional, controlled vs. uncontrolled, and randomly-assigned). They must also understand the simplest ways of taking related factors into account (comparing subgroups, using averages, comparisons, ratios, comparisons of ratios and relative risks), and the more complex ways of taking related factors into account (using weighted averages to adjust rates and percentages for the influence of a binary confounder). Statistically literate adults must understand association-generated measures involving epidemiological models: the percentage – and number – of cases that are attributable to an associated factor. They must also be able to describe and compare rates and percentages presented in tables and graphs. Here are some right/wrong questions involving context:

1. A controlled study means it is an experiment – not an observational study. [Answer: False. Explanation: A controlled study is any study having more than one group.]
2. Do these statements have the same meaning? (A) Widows are more likely among suicides than widowers. (B) Widows are more likely to commit suicide than widowers. [Answer: No. Explanation: Suicide is the common-whole in A, the common-part in B.]

3. What percentage of low-weight births are attributable to the mother smoking? In the US in 2002, the percentage of newborns that have low birth-weight is 12.2% among mothers who smoke and 7.5% among non-smoking mothers. See Figure 11.1 (Table 84, 2004 US Statistical Abstract).

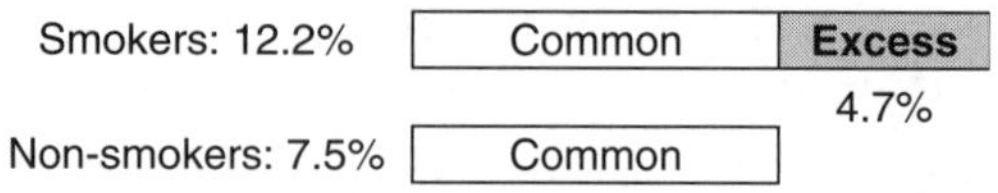

Figure 11.1 Percentage of newborns with low birth-weights.

[Answer: 39%. Explanation: The risk 'attributable to' membership in the group with the larger rate (the 'attributable risk' or 'excess risk') is the rate difference (the 4.7 percentage point excess in the larger) divided by the larger rate (12.2%) shown as a percentage: 100%(12.2%-7.5%)/12.2%.]

4. In 2002, of the low-weight births to US mothers who smoke, how many are attributable to their mother smoking? Of the 4.02 million US births in 2002, 8.0% were low weight so that 10% of these mothers reported smoking. [Answer: ~19,000. Explanation: 400,000 (10%) babies born to mothers who smoke. 48,000 (12%) have low birth-weights. 19,000 (39%) are attributed to their mother smoking.]

5. At a given hospital, suppose that entering patients are in either good condition or poor condition and that the patient death rate is 4% for patients in good condition versus 12% for those in poor condition. What is the average death rate for this hospital if 75% of the patients are in poor condition? [Answer: 10%. Explanation. This average is a weighted average – not a simple average. $0.75*0.12 + 0.25*0.08 = 0.08 + 0.02 = 0.10$.]

6. If one of these percentages is bigger which is it? (A) The percentage of infant deaths which are due to birth defects. (B) The percentage of infants who die due to birth defects. [Answer: A. Explanation: Let X be 'due to birth defects', Y be 'death' and Z be 'infant'. A = P(X|YZ) and B = P(XY|Z). Since XY $\leq$ X and YZ $\leq$ Z, it follows that P(XY|Z) $\leq$ P(X|YZ).]

7. What is the chance that a young adult who fails to graduate from high school will spend time in prison? Suppose that 72% of the young adults in prison did not graduate from high school whereas 12% of all young adults did not graduate from high school. Suppose that 5% of all young adults spend time in prison. [Answer: 30%. Explanation: Let X be 'did not graduate from high school' and Y be 'in prison'. P(X|Y) = 72%; P(X) = 12%, P(Y) = 5%. Bayes Rule: P(X|Y)/P(X) = P(Y|X)/P(Y). So P(Y|X) = P(X|Y)[P(Y)/P(X)] = 0.72[.05/.12] = 0.3 = 30%.]

8. Do these statements say the same thing? (A) Smoking is twice as prevalent among women as among men. (B) Women are twice as likely to be smokers as are men. [Answer: Yes.]

9. Do these statements say the same thing? (A) Smoking is 50% more prevalent among women than among men. (B) Men are 50% less likely to be smokers than are women. [Answer: No. '50% more' is equivalent to '33% less'.]
   A statistically literate person should be able to envision what it means to 'control for' or 'take into account'. In the simplest case, it means recognising that a difference in ratios may be spurious after controlling for a related factor – a confounder.

10. Is this difference in state NAEP scores shown in Table 11.2 real or spurious?

Table 11.2 NAEP Mathematics Scores 2000, grade 8: MD vs AZ.

Average Score	State	Internet access at home?				
		YES		NO		ALL
274	Maryland (MD)	281	70%	258	30%	100%
271	Arizona (AZ)	281	55%	258	45%	100%

Source: US Dept of Education, National Assessment of Educational Progress (NAEP).

Since the three-point difference in scores between states vanishes for each of the two subgroups, a statistically literate student must recognise the original difference by state is spurious after taking into account the different mixtures of students having Internet access at home.

11.10 Assessing statistical literacy: Assembly

Recall that 'assembly' means the choices in defining, selecting or presenting statistical relationships. Consider these examples of 'assembly' from Schield (2007):

- OPEC countries supply 50% of US oil imports, but only 30% of US oil usage.
- The average US farm is 440 acres; the average US family farm is 326 acres.
- Annual income is $43K for households, $53K for families and $62K for married couples.
- In 2005, the world gained 2.3 people per second (over 74 million people per year).

To understand the influence of assembly on a statistic, statistically literate consumers should know that as the definition of a group becomes more restrictive, the size of the group will decrease, but that the size of a ratio involving that group as the whole may increase. They should know that arithmetic differences are typically bigger than percent differences or times ratios when comparing large numbers but that percentage differences or times ratios are typically bigger than arithmetic differences when comparing small numbers. They should know that the choice of the cut point in forming subgroups from a continuous distribution can strongly influence the difference and ratios of statistics for the subgroups.

The following items assess one's knowledge about the influence of 'assembly' on a statistic.

Example 1: Which definition gives the larger number?

1.1 Number of teens: 'those 13–6' vs. 'those 13–19'. [Answer: The less restrictive group (13–19).]

1.2 Smokers: those who smoked in the last month vs. in the last year. [Answer: Longer time period.]

1.3 Average incomes: those ages 20–65 vs. those ages 1–100. [Answer: Ages 20–65.]

Table 11.3 US Women (in millions) who had a child in 2004 by family income.

<10K	10–19.9K	20–24.9K	25–29.9K	30–34.9K	35–49.9K	50–74.9K	75K and up
4.2	6.2	3.4	3.8	3.6	8.9	10.6	12.5

Source: 2006 US Statistical Abstract, Table 88.

Example 2: Who had more babies: rich mothers or poor mothers? (see Table 11.3)

2.1 Define 'Rich' as 35K and up; define 'Poor' as under 35K. [Answer: Rich mothers.]

2.2 Define 'Rich' as 75K and up; define 'Poor' as under 25K. [Answer: Poor mothers.]

Table 11.4 Estimated US persons (thousands) living with AIDS by race/ethnicity for 2003.

ALL	Non-hispanic white	Non-hispanic black	Hispanic	Other
406	147	172	80	5

Source: 2006 US Statistical Abstract, Table 180.

Example 3: According to Table 11.4, which group had the most cases of AIDS? (Assume that 75% of Hispanics are white.)

3.1a. Non-Hispanic white	b. Non-Hispanic black	[B]
3.2a. White (including white Hispanics)	b. Black (including black Hispanic)	[A]

Allocating Hispanics by race gives 207,000 whites and 192,000 blacks.

Example 4: Which saves more gas in going 10,000 miles?

4.1 Improving car A's miles per gallon from 10 to 20, or

4.2 Improving car B's miles per gallon from 20 to 100? [Answer: 4.1 saves 500 gallons – half of the initial 1,000; 4.2 saves only 400 gallons – 80% of the initial 500.]

11.11 Assessing statistical literacy: Randomness

Even before learning anything mathematical about chance, a statistically literate person should recognise the law of Very-Large Numbers: the unlikely is almost certain, given enough tries. More specifically, a random event with 1 chance in N of occurring on the next try is expected – is more likely than not to occur at least once – in the next N tries. (Schield, 2005).

Even before they can calculate a margin of error, a statistically-literate person should recognise that statistical significance can be determined, given two sample means and their associated margins of error or confidence intervals. If two 95% confidence intervals do not overlap, that test indicates the difference between these sample means is statistically significant at the 5% level. When they do overlap, that test indicates the difference is not statistically significant. A more accurate test (e.g. a t-test) will find that some of these cases of statistical insignificance are actually statistically significant. But a more sophisticated test (resampling) may find that some of the t-test cases of statistical insignificance are actually significant. Statistical significance is determined by the test.

1. When an activity involves discrete outcomes with various probabilities, is the expected value always one of the outcomes? [Answer: No. An average need not be one of the outcomes.]

2. When flipping a fair coin, what is the chance that the next 5 flips are all heads? [Answer: 1 in 32.]

3. When flipping a set of five coins 32 times where a success is flipping all five heads, what is the expected value: the number of successes one can expect? [Answer: One.]

4. If the expected value is a possible outcome, does this mean that expected outcome is more likely than not? [Answer: No. When flipping a fair coin four times, the expected outcome is two heads. This is the most likely outcome, but the chance of two heads is less than 50%.]

5. If the expected value is a possible outcome, does this mean the expected value is the most likely outcome? [Answer: No. If the probability distribution for discrete outcomes is symmetric with a bowl shape, the expected value will be the middle which has the smallest probability.]

6. Is a rare coincidence – an event that is extremely unlikely if due to chance – therefore highly unlikely to be due to chance and thus highly likely to be due to some causal factor? [Answer: Not necessarily. If predicted before the fact, probably. If selected after the fact, probably not.]

7. Suppose that in a two-candidate race, a poll indicates that one candidate has 55% of the vote while the other has the remainder. The 95% margin of error is 4%. Is this difference statistically significant when using the simple overlapping confidence-intervals test? [Answer: Yes.]

8. Suppose that a survey found that the average income was $55,000 for one group – 10% more than that of a second group. The 95% margin of error for both groups was $3,000. Is this difference statistically significant when using the simple overlapping confidence-intervals test? [Answer: No. The average income for the second group must have been $50,000. With a $3,000 margin of error on each value, the $5,000 difference is not statistically significant.]

9. Suppose that subjects are randomly assigned to two groups and that the difference in their outcomes is statistically significant. Can we expect that taking into account a pre-existing condition will make the difference statistically-insignificant? [Answer: No, random assignment tends to allocate any pre-existing condition equally to the two groups, so taking into account this condition is not expected to change anything.]

10. If subjects are assigned to two groups without random assignment and the difference in their outcomes is statistically significant, can taking into account a related condition give a new difference that is statistically insignificant? [Answer: Yes if strong enough. See Schield (2004b).]

11.12 Assessing statistical literacy: Error/bias

1. Will having a larger sample mitigate the influence of error or bias? [Answer: Not generally.]

2. Can getting a larger sample (conducting a census) increase the chance of error? [Answer: Yes.]

3. Of those surveyed, 20% did not respond. Is this non-response bias? [Answer: No. Non-response causes bias only if non-respondents would have answered differently than respondents.]

11.13 Assessing the influence of confounding

Ridgway *et al.* (2008: 1) note that 'most interesting problems are multivariate', and so 'the curriculum (and ideas about statistical literacy) should encompass reasoning with multivariate data'. Statistical educators may question whether it is possible teach students about the influence of a confounder on a statistic without teaching multivariate regression and the associated diagnostics and assumptions. Schield (2006) demonstrates a simple graphical technique for a binary predictor and a binary confounder that bypasses the need to discuss the assumptions of linear regression. This graphical technique involves weighted averages and uses a statistical principle from the 1960s called 'standardising'.

The following exercise uses this technique to show the influence of a confounder on three things: the size of an association; the number of cases attributable to a related factor; and the statistical significance of a difference between two groups.

11.13.1 The size of an association

To see the influence of a confounder on an association, consider two hospitals: Rural and City. Patients in good condition can walk in; patients in poor condition are carried in. Suppose the death rates (hypothetical) are 2% and 7% for those in good and poor condition at the Rural hospital; 1% and 6% respectively at the City hospital. Suppose that 90% of the City patients are in poor condition (30% of the Rural).

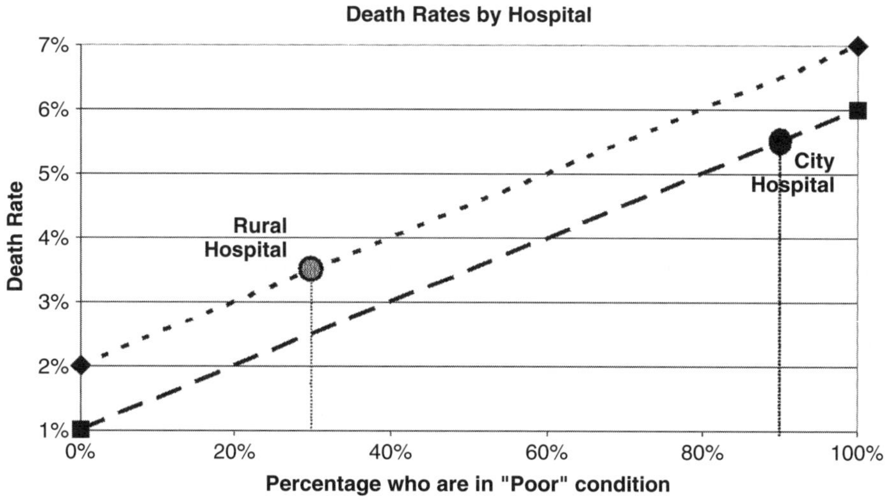

Figure 11.2 Raw hospital death rates.

1. What are the average deaths rates at these two hospitals? [Answer: City (5.5%), Rural (3.5%). Algebraic solution for a weighted average: City (0.9*0.06 + 0.1*0.01), Rural (0.3*0.07 + 0.7*0.02). Figure 11.2 shows how this weighted average can be solved graphically for each hospital.]

2. Which hospital has the higher death rate after taking into account the difference in patient mix? Figure 11.3 illustrates standardising: taking into account a difference in mix. Suppose the combined hospitals have 60% of their patients in poor conditions. If we standardise on 60%, we see that Rural has a higher standardised deaths rate (5%) than City (4%). This reversal is an example of Simpson's Paradox. This simple graphical technique allows students to work problems and thereby calculate the influence of a binary confounder on an association.

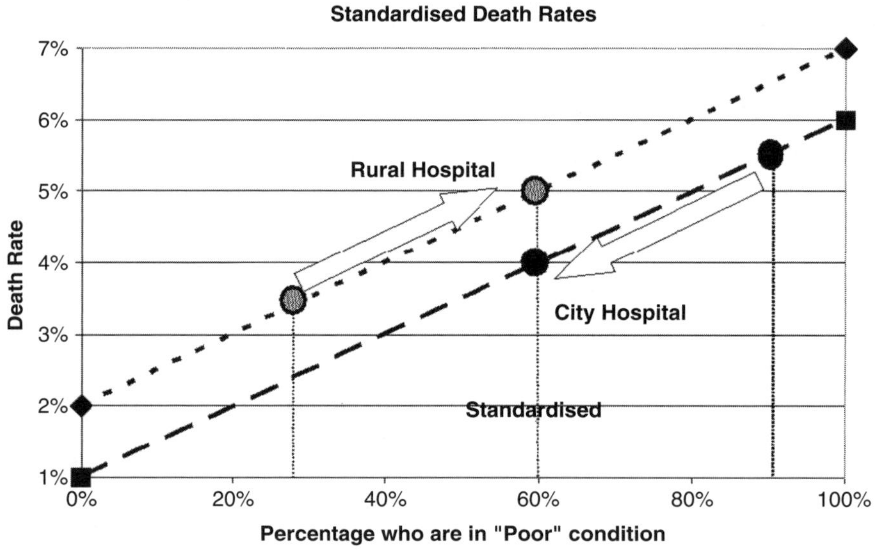

Figure 11.3 Standardised hospital death rates.

11.13.2 The number of cases attributable to a related factor

Confounding can influence speculative statistics based on epidemiological models. Suppose that among mothers who do not smoke the percentage of their babies who have low birth-weights is 6% and 11% among older and younger mothers; 11% and 16% among those who do smoke. Suppose that those under 19 are 10% of non-smokers (50% of the smokers).

1. Among non-smoking mothers, what percentage of babies have low birth-weights? [Answer: This problem in weighted averages can be solved

algebraically. Among non-smoking mothers: 0.10*0.11 + 0.90*0.06 = 0.065. Among smoking mothers: 0.5*0.16 + 0.5*0.11 = 0.135. Or it can be solved graphically as shown in Figure 11.4. Either way, the percentage is 6.5% among non-smoking mothers and 13.5% among mothers who smoke.]

2. What percentage of low birth-weight babies with mothers who smoke are attributable to their mother smoking? [Answer: 52%: (13.5%–6.5%)/ 13.5%. See prior discussion on excess risk.]

3. How many babies having low birth-weights are attributable to their mother smoking? Assume there were 3.5 million births. Assume 25% of these mothers smoked. Of the 875,000 babies whose mothers smoked, 13.5% (118,125) have low birth-weights. Of these 118,125 low birth-weight babies whose mothers smoked, 52% (61,250) are attributable to their mother smoking. [Answer: 61,250.]

4. After taking into account the influence of age, what are the standardised percentages of babies who have low birth-weights? Assume that 20% of all mothers are 19 or younger. [Answer: The standardised percentage of babies who have low birth-weight is 7.0% among non-smoking mothers, 12.0% among mothers who smoke. Algebraically: 0.2*0.11 + 0.8*0.6 = 7%; 0.2*0.16 + 0.8*0.11 = 12%. Figure 11.4 illustrates these standardised values graphically.]

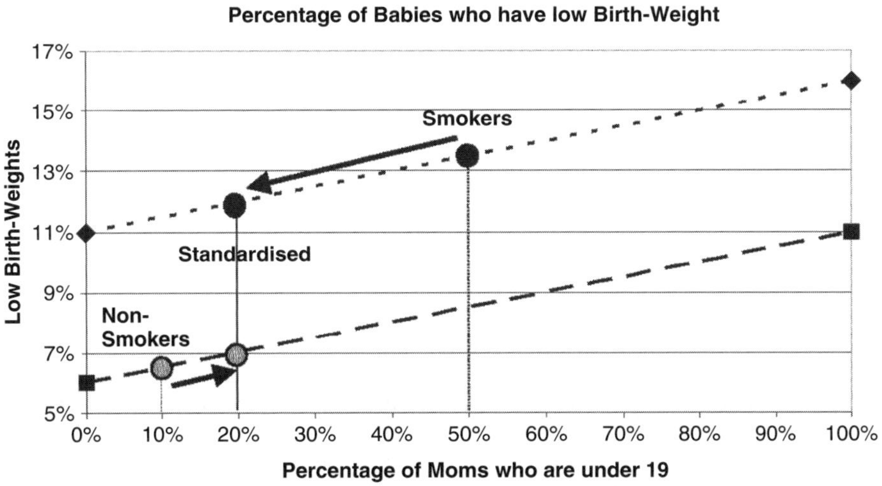

Figure 11.4 Influence of confounding on percentages and cases attributed.

5. Using the standardised values what percentage of low birth-weight babies whose mothers smoke are attributable to their mother being a smoker? [Answer: 42%: (12%–7%)/ 12%.]

6. After taking into account the influence of age, how many babies having low birth-weights are attributable to their mother smoking? Assume 3.5 million. Assume 25% of these mothers smoked. Of the 875,000 babies whose mothers smoked, 12% (105,000) have low birth-weights. Of these 105,000 low birth-weight babies whose mothers smoked, 42% (43,750) are attributable to their mother smoking. [Answer: 43,750.]

7. Compare the number of babies who have low birth-weights that are attributable to their mother smoking before and after taking into account the influence of age. In both cases, there were 875,000 babies whose mothers smoked.

   - Without taking age into account, 118,125 had low birth-weights. Of these 61,250 (52%) were attributable to their mother smoking.
   - After taking age into account, 105,000 had low birth-weights. Of these 43,750 (42%) were attributable to their mother smoking.

8. Analyse the difference in these two cases. Taking age into account reduced the number of low-weight births attributable to smoking by almost 30% – from 61,250 to 43,750. Figure 11.5 illustrates these differences.

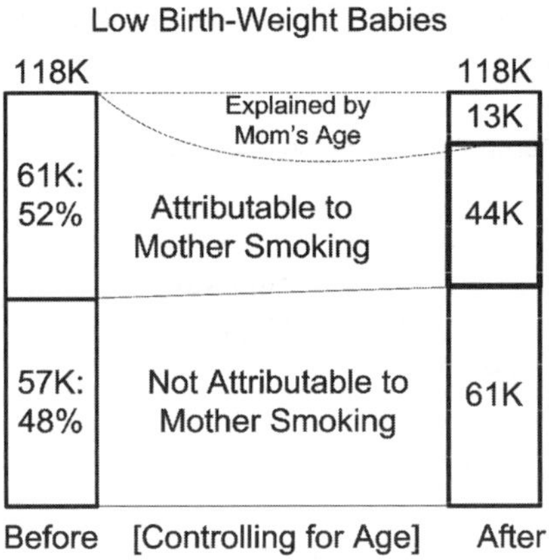

Figure 11.5 Low birth-weights attributed to smoking: influence of age.

Figure 11.5 preserves the original number of cases. Statistical educators can decide how best to make and explain these comparisons. Controlling for age decreased the number of low birth-weight births attributed to smoking from 61K to 44K – a reduction of almost 30%.

11.13.3 The statistical significance of a difference between two groups

Controlling for a confounder can influence whether a difference that is statistically significant becomes statistically insignificant – or vice versa.

1. Are these differences statistically significant if the 95% margin of error is three percentage points? Using the gap between confidence intervals as a simple – but conservative – test for statistical significance, the initial seven-point gap (13.5% vs. 6.5%) is statistically significant.

2. Are these differences statistically significant after controlling for the influence of age? [Answer: No. The standardised gap is five points: (12% vs. 7%). Using the simple overlapping confounder intervals test, this difference is not statistically significant.
Both of these results can be seen graphically in Figure 11.6.

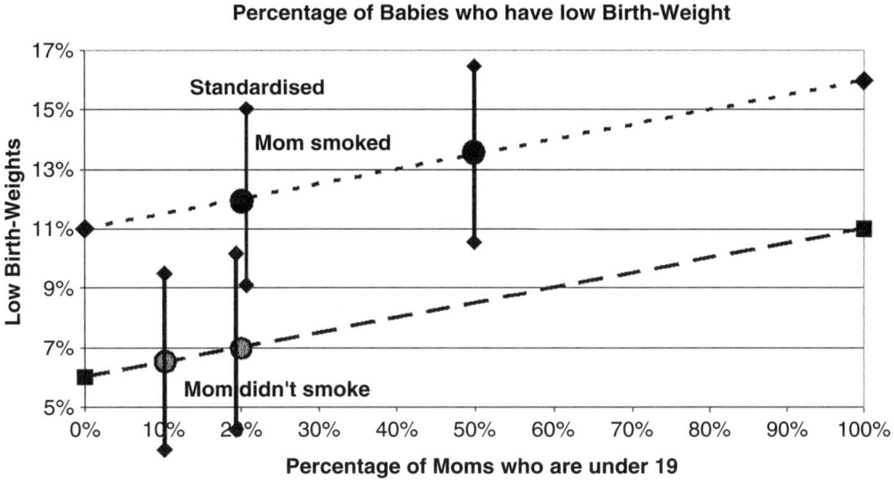

Figure 11.6 Influence of confounding on statistical significance.

11.14 Student feedback

Student feedback indicates they have found these exercises very helpful in understanding the key idea that association is not causation. Their primary request is for a wider variety of problems so they can strengthen their skills. Students are amazed at how the choice of a definition can so easily influence the size of a number. They are astounded that an arithmetic association can change direction after taking something else into account. Students are dismayed that risks 'attributed to' or 'due to' an associated factor are really speculative statistics that involve no causal claim. They are very dismayed to learn that these excess risk numbers can change after controlling for a related factor. And they are often bewildered when they learn that a statistically significant difference can become statistically insignificant (and vice versa) after controlling for a confounder. In summary they are amazed that statistics are so easily influenced by social construction, whereas the numbers in arithmetic are immune. As a result, they often say, 'I see statistics differently now'. They strongly agree that one must definitely 'take care' when using statistical associations as evidence for causal connections in arguments.

11.15 Assessment

These exercises provide a basis for assessing student knowledge of the various influences on a statistic. Assessing content knowledge is important, but assessing student attitudes is more important. Macnaughton (2004) argues that the primary goal of an introductory statistics course should be to give students 'a lasting appreciation for the value of statistics'. This may be difficult and not as important in teaching statistical competence when students have no idea of how valuable statistical inference is in certain situations. This should be easier and more important in teaching statistical literacy where students can readily find examples of statistical illiteracy in the everyday news.

11.16 Conclusion

Statistical literacy and statistical competence are related but different. Neither guarantees the other. Students in non-quantitative majors need statistical literacy. Students in some quantitative majors may need both.

A statistical literacy course should be designed to satisfy the needs of citizens in a modern, data-driven society, to help them think critically about statistics when used as evidence in arguments. There are many ways to design a course to achieve these goals. However, if a course is to carry a statistical literacy designation and meet these goals it should: (1) study all sources of influence on a statistic; (2) choose topics based – in large part – on their prevalence in the everyday media; and (3) inspire data consumers to see a positive value in the material presented.

Statistical literacy courses are becoming increasingly common in US four-year colleges. This increase allows the 40% of US college students in non-quantitative majors a better chance to think critically about numbers in the news. With empirical data on statistics in the media, with a distinct content and with new forms of assessment, statistical literacy is rapidly emerging as a new course in the liberal arts.

Part C

ASSESSMENT USING REAL-WORLD PROBLEMS

12

Relating assessment to the real world

Penelope Bidgood

12.1 Introduction

Modern statistical education, at all levels, places increasing emphasis on students' abilities to think and reason statistically using real data in appropriate contexts. Rather than focusing on statistical skills, procedures and computations, there is a growing call to encourage students to be statistically literate (Ben-Zvi and Garfield, 2004a; Rossman and Chance, 2002). Thus, there is an increasing emphasis on data collection, exploration of data and the interpretation of results, which are dependent upon context. Snee (1993) argues that we can help students better learn statistical thinking and methods by focusing the content and delivery of statistical education on how people use statistical thinking in real-life situations.

Whilst statistical reasoning and thinking are not always clearly defined, Wild and Pfannkuch (1999) develop a model that provides some clarity and gives a way of examining what these terms mean. Their five components of statistical thinking are: recognition of a need for data; ability to 'transnumerate' the data; recognition of variation; being able to reason from models; and being able to integrate statistical and contextual knowledge. Assessment of statistical reasoning and thinking presents its own challenges.

Here, two aspects in relating assessment to the real world are considered – first, to use assessment to investigate real and relevant data

Assessment Methods in Statistical Education Edited by Penelope Bidgood, Neville Hunt and Flavia Jolliffe

and second, to prepare students for their future working lives in being able to analyse data appropriately and to communicate their findings effectively. These each reflect the 'Increasing emphasis and research in the statistical and statistical education community on data driven educational strategies; on the nature of statistical thinking and reasoning and on the roles of technology in statistical education' (MacGillivray, 2004: 233). The two aspects of assessment described above are clearly related to Wild and Pfannkuch's model, but a more comprehensive consideration of assessment of statistical literacy and thinking per se can be found in Chapter 6 by Jolliffe.

Business, psychology, health and medicine, economics, geography and most sciences are some of the disciplines that incorporate statistics into their courses. Budgett and Pfannkuch in Chapter 9 describe statistics assessment with budding journalists, lawyers, politicians and sociologists. Whatever the main discipline, it has been found that, in such courses, students respond better to statistics if the data are relevant to and applicable in, the main subject area (Garfield, 1995; Jolliffe, 2007; Griffiths and Sheppard, Chapter 4).

The ASA funded the GAISE Project, which focused on introductory college courses. The project report (ASA, 2007) recommended that the teaching of introductory statistics should emphasise statistical literacy and develop statistical thinking; use real data; stress conceptual understanding rather than mere knowledge of procedures; foster active learning in the classroom; use technology for developing conceptual understanding and analysing data; use assessments to improve and evaluate student learning.

Likewise, a report on the state of mathematics, statistics and operational research (MSOR) at 71 tertiary institutions in England and Northern Ireland in 2000 reported that, 'Student engagement and performance has often been greatest when dealing with well-focused problems of a practical nature' (MSOR Overview Report, 2000). It also noted that assessment emerged as one of the most problematic areas in MSOR provision and, in a follow-up project, it was reported that good practice involved a wide range of assessment instruments to be used to address learning outcomes (Bidgood and Cox, 2002). There are varied assessment strategies in statistics, whether in specialist or service courses. Different methods of assessment are appropriate for different elements of the curriculum and for different students – statistics is widely used in many fields and there are differences in the approach to be taken in service and specialist courses.

Each of these aspects of real-world assessment has been enhanced by the huge advances in technology over the last few years. The Internet allows access to a vast range of data from many fields, thus creating a valuable resource for lecturers and students. Statistical software has greatly expanded the range of analyses that students can conduct and this affects the assessment process. Using software that allows students to visualise and interact with data appears to improve students' understanding of random phenomena and their learning of data analysis (Garfield, 1995).

12.2 Assessment with real data

Data that are real and interesting motivate students – they want to use material which relates to them and/or their discipline and this can help them enjoy learning statistics. Consequently the teaching, learning and assessment resources produced should use authentic data and relevant scenarios.

There are many web sites that, taken together, allow thousands of data sets to be downloaded, so that in theory, lecturers and students have access to a rich source of information. For example, the MSOR web site (www.mathstore.ac.uk) lists and comments on numerous sites where data can be found. However, many of these data sets lack descriptions of the contexts in which the data were obtained and some of the databases are difficult to access. Also, there are very few web sites that provide help with teaching and learning activities using the data set and not every lecturer has sufficient time to create examples in context. For example, Barnett (2004) reported that there was not a vast amount of UK-based university-level statistics teaching material freely available on the Internet for general use, as popularly supposed. Mashhoudy, in Chapter 21, notes the shortage of real, accessible data that can be used in more advanced statistics, such as multiple regression techniques and log linear modelling.

The massive expansion in technology means that large data sets, when available, can be used and stored easily. Further, requiring students to perform real statistical analysis on real data and to report their results is now practicable, as lecturers are freed from using limited examples, due to computational restrictions; more varied types of statistical methods can be used.

> For example, technology use has allowed earlier accessibility to complex investigations, exploratory data analysis and visualisation, simulation and re-sampling'. (Begg *et al.*, 2004: 276).

In service teaching for non-specialists, classes are often very large, with more than 500 students being quite typical in the business field. Large numbers of students, with the associated plagiarism issues, has an effect on the types of assessment used, the way in which they are marked and feedback to students. The Plagiarism in Statistics Assessment (PiSA) project (Bidgood *et al.*, 2007) found that, to counteract these problems, there was an increasing move from take-home assignments towards 'hands-on' practical assessment, with lecturers keen to use real and relevant data. Students could be given some data in advance and then answer questions, either by hand or on a computer, in a later, timed, supervised session. This allows students to discuss the data with each other, but ensures that what is handed in and marked is their own work. A common technique is to ask students to gather their own data, either on themselves, from experiments or through general or specialised web sites, with possible restrictions on the type and amount of data collected. The use of real data to motivate assessment is shown by a number of contributors to *Assessment Methods in Statistical Education*; see, for example, Griffiths and Sheppard (Chapter 4), who use a human biology database, and Lopez *et al.* (Chapter 14), who take examples from agriculture. Both Tyrrell (Chapter 13) and Hannigan (Chapter 15) describe surveys conducted by students on other students. While there are risks that students might find data for which the analysis has already been done and is published, there are mechanisms to avoid this sort of plagiarism, as reported by the PiSA project.

One recent development to address such issues was the STARS (Creating Statistical Resources from Real Datasets) project. Its main aims were to make available real data sets and associated scenarios applicable in the subjects of psychology, health and business and to develop learning and assessment materials to accompany these data sets for use with various packages (stars.ac.uk). Two assessment tools were also developed as part of the project, which each allow individualised data sets to be produced, together with assignments and solutions.

Wild *et al.* (1997) use some traditional methods of assessment, such as multiple-choice tests, on real data and stories, to measure students' statistical reasoning skills.

While many teaching, learning and assessment materials using real life data have been produced for introductory service courses, recently the focus has been on the more specialist programmes, where students may be expected to have stronger mathematical backgrounds. Rossman and Chance (2002) describe the development of curricular materials for such students and their stated principles include motivating students with real data and problems, fostering active explorations and developing problem-solving skills. They found that data from scientific studies, popular media or student-collected motivate the students, who are thus introduced to statistical concepts, methods, and theory through a data-oriented, active learning pedagogical approach. Tyrrell describes assessment in a class with both specialist statistics and business students (see Chapter 13). Mashhoudy describes how both elementary investigations, through descriptive statistics and graphical methods, or more complex modelling, can be applied to real and constantly changing data which students collect from the Internet (see Chapter 21).

12.3 Assessment to help employability

One aspect of relating assessment in statistics to the real world is to develop the statistical skills and ability to reason statistically that will help students in their future workplace. The growth of awareness of statistical literacy and the consequent importance of statistical thinking and reasoning has led to changes in statistical teaching at the tertiary level. This reflects, in part, demands from employers who want graduates with a balance of technical statistical skills, including analysis and interpretation, with the ability to communicate their findings. The aim is to produce graduates who can carry out a statistical investigation, choosing appropriate methods to analyse their data effectively and presenting their results successfully. More advanced students should be able to carry out the modelling process, from understanding the formulation of the problem, to communicating their results, whilst appreciating the limitations of their methods.

There are many modes of assessment that help students to prepare for their future working life, for example computer-based assignments, investigations, modelling assignments, presentations and group work. At one time statistics was usually assessed mainly by formal examination; as the applications in the subject have become more widely taught, with an increasing emphasis on statistical literacy and reasoning, and with technological advances, the subject has lent itself to a more varied assessment regime. Students can be set more realistic problems; they can complete weekly online quizzes; they can carry out simple experiments and simulations; they can keep portfolios of their work; they are often required to communicate the results of their analysis graphically, verbally or in writing, including poster presentations; they can be asked to critique the study designs and analysis of others.

An example of the latter kind of assessment is given in Chapter 10, where McNiece explains a research-based assessment in a medical statistics module, where students have to find their own example of a medical case study. There are also arguments for requiring graduates to be able to write about their findings, not least of which is the ability to communicate with non-statisticians. Forster and Wild (see Chapter 8) describe assessment of students' writing about statistics.

Many of the different types of assessment occur in both case-studies and projects. Typically, as the case- study is assessed within a module, covering a particular area of statistics, such as regression modelling or time series analysis, the choice of techniques or models to be used is limited. Nevertheless, the task of carrying out an extensive study and communicating the results is good preparation for future working life. A project, on the other hand, in UK terms, is typically a sustained, usually individual, piece of work, taken in the final year of degree study, where a student may work on almost any area of statistics. Case studies and projects, whether carried out in groups or individually, are good assessment tools to measure students' conceptual understanding and their ability to think statistically, as well as their technical skills, both statistical and transferable, such as communicating results and using appropriate technology.

Group work, which is often used in case studies, has the advantage of mimicking real life in the sense that often, professional statisticians are working in teams. The PiSA project (Bidgood *et al.*, 2007) reported that in most cases the groups were small, usually with fewer than four students working together, to ensure that everyone made a valid contribution to the study. Typically, this work is assessed by a written and/or oral report, although increasingly, poster presentations are used. Normally, for the latter, each individual must be present at the poster session to answer questions on the work. Posters are helpful in preparing those students who go onto further academic work, as such displays are often part of academic conference proceedings. In some cases, although students work together, they produce individual reports or answer individual questions on the work, which avoids anyone freeloading and not contributing fully.

The problems tackled in an individual or group project may be of a more open-ended nature, allowing students to increase their knowledge of statistics by studying a topic in greater depth and/or by applying techniques learned in a new situation. 'Project work is a method of allowing students to make use of what they have learned in statistics classes in a practical context. It is this practical application of projects that make them such a useful part of the learning process' (Starkings, 1997: 139). The broad aims of the project are twofold – to give students the experience of undertaking some personal research and/or scholarship in a branch of statistics, and to develop their independent learning and other key skills. Holmes (1997) notes that one of the main aims for doing projects is for students to become autonomous learners and deems this worthwhile in itself. Although he is writing about projects for school children, the same may be said of university and college students.

At the author's own university, the learning outcomes for the individual project are that: students should be able to carry out a literature search; devise and write a concise plan of a proposed research project or dissertation; undertake an investigation of the planned topic and compare the outcomes with the original proposal; produce a well-structured written report; give an oral presentation or answer questions clearly and concisely in a structured interview about their work; demonstrate that they can use IT skills; and demonstrate their knowledge of computing skills if appropriate.

Although assessing project work can be complex, due to its diverse nature, there is usually some recognition of development and progress, as well as the final report. As such the assessment will place a greater emphasis on ability to plan work, manage time effectively and research background information. Again, at the author's own university, students have to keep monthly logs of their work and produce an interim report and presentation at approximately the half-way stage. Although these components account for only 20% of the final mark, they do enable students to respond to the feedback they receive, before completion of the final report and oral examination.

12.4 Concluding remarks

There are two aspects in relating assessment to the real world – first, to use real data in context and secondly, to imitate some part of a professional or academic statistician's working life. Although each aspect can be found in either service or specialist courses, the former is perhaps more relevant to service courses where it is important to choose examples of interest, applied in the main subject area, and the latter more pertinent in specialist courses which include the modelling process, case studies and projects. This is summarised by MacGillivray (2004: 233): 'good data analysis practice and thinking require a balance of technical statistical skills, quantitative skills, judgement, the ability to comprehend and model (in the mathematical sense) in contexts that may be unfamiliar, and analysis and interpretation that include balance, synthesis and communication'.

The use of real data helps undergraduate students in service courses as they do not always see the immediate relevance of statistics course in their own discipline. This is greatly enhanced by developments in technology, which enable students to access various data sets, or to store data they have collected efficiently. Service courses in statistics are typically in the first year of a degree programme, although students often require statistical skills and reasoning in later parts of their programme, particularly when they have to analyse data in their own final year projects. The aim should be to produce graduates, whether specialist statisticians or not, who are familiar with the different phases of a statistical investigation, are able to choose correct methods and models in order to analyse data effectively, and who can present their results to both statisticians and non-statisticians.

Assessment that relates to the real world covers many aspects of what is required of the modern undergraduate – to access and work with genuine data; to work collaboratively, to be engaged in active learning, to have conceptual understanding, to be able to use technology appropriately, and to be able to communicate statistics effectively.

13

Staged assessment: A small-scale sample survey

Sidney Tyrrell

13.1 Introduction

This chapter discusses a staged assessment strategy in the context of teaching students with a limited statistical background both the theoretical and practical aspects of conducting a small-scale sample survey.

The problem of balancing the theory and concepts of sampling with the more practical aspects of carrying out a survey has long been recognised. O'Muircheartaigh (2005: 4) writes of the challenge facing teachers of sampling in combining the technical aspects, mathematical and statistical, with the practical. His view is that:

> The ideal sampler should have a technical understanding of sample design and estimation . . . and an understanding of the interdependence of sample design and survey operations.

Both Statistics Canada (Gambino and Gough, 2005) and the UK Office for National Statistics (ONS) (Brown, 2007) recognise the problems even for students with a statistics background. The University of Southampton's Masters course in Official Statistics is linked with ONS, providing a practical demonstration of how knowledge of theory has to be balanced with the requirements

Assessment Methods in Statistical Education Edited by Penelope Bidgood, Neville Hunt and Flavia Jolliffe

of the practicalities and problems of actually carrying out surveys. Brown (2007: 5) writes:

> By putting both together we aim to ensure that the learning of theory is integrated with its practical application and not seen by students as an irrelevant and abstract mathematical exercise.

In the UK at undergraduate level there has been a call by the Economic and Social Research Council (ESRC, 2006) to address the perceived need for the development of undergraduate curricula in quantitative methods. Of particular relevance to the teaching under discussion were the need to develop curricula which, firstly, show students that in building on their school experience, and their IT practice, they already have the skills needed for the foundations of quantitative research, and secondly which encourage students to conduct their own surveys and to analyse the results as part of their course work (Gibson *et al.*, 2007).

Although applying particularly to social scientists this has relevance for business and IT students as well. Every undergraduate student needs to acquire professional skills as part of their university education and every professional statistician needs to understand the power and the pitfalls of survey techniques. Petocz and Reid (2007: 7) argue that, ‘In a statistics course, we have the opportunity to support students’ professional formation by incorporating professional skills and dispositions into the curriculum and the assessment, and in return we benefit from students’ increased engagement’.

The case therefore appears to have been made by others for a combination of theory and practice in this field, both for students from a non-specifically statistical background as well as those with one.

At the time this assessment was carried out students took eight modules a year for their degree, of which one or two were free-choice options. The module concerned, ‘Data Analysis’, was offered to second years, some of whom came from the Business School – some IT students, as well as those studying mathematics and statistics. There was, however, a requirement that they had undertaken a module covering basic statistics. In practice this knowledge proved to be very basic.

The module attracted 40 students and was taught over 22 weeks in a weekly two-hour slot with access to PCs for one of those hours if required. It aimed to introduce students to data gathering, manipulation and validation, looking at the major users and uses of data in the real world. Half of the course was concerned with the construction and use of databases, and in the other 11 weeks students were introduced to SPSS and given the experience of all stages of a statistical investigation.

The intended learning outcomes being partly or completely assessed were that the student should be able to:

- Describe different methods of data collection.
- Retrieve data from a variety of sources.

- Manipulate, edit, and validate data presented in a variety of formats.
- Design, conduct, analyse, report and critically evaluate a small-scale sample survey.

The assessment consisted of three parts:

- A database exercise 40%
- An in-class test on the use of SPSS 20%
- The survey 40%

The teaching was divided between the author, who was also the module leader, and a colleague who led on databases. The 22 weeks were arranged into blocks of four or five weeks, alternating database work with survey work, which, as described later, greatly assisted in the delivery of the teaching.

13.2 Assessing for learning

Petocz and Reid (2007: 8) point out 'Using assessment to support learning, and in particular to help students develop their professional dispositions, is an important challenge for a statistics course – particularly a first service statistics course'. It was Ramsden (1992) who wrote 'assessment drives learning', and it has long been the author's view that thinking about the appropriate assessment for learning outcomes is the major influence in effective teaching and student engagement in their learning. Previous experience undertaking survey work with students had been disappointing. Students had worked in small groups, but in some the difficulties of the group dynamics dominated the learning experience, and in others there appeared to be no aspiration to anything approaching a professional standard of work. The organisation of the teaching plan had meant that they were given the theory then left working on their own for too long to effectively put it into practice. Feedback, throughout the assessment, was clearly an important element to consider if one takes the constructivist view of learning that students actively construct their learning, and that it is an adaptive process.

'An assessment activity can help learning if it provides information to be used as feedback, by teachers, and by their students, in assessing themselves and each other, to modify the teaching and learning activities in which they are engaged'.(Black, 2005, Slide 6).

It was clear to the author that more was needed in terms of getting the students to think more deeply about the task, to think collaboratively, and to raise their aspirations. They needed to take responsibility for their own learning, but with appropriate stimulus and encouragement. The author needed to engineer a situation that enabled learning rather than deliver teaching, and in the context of analysing primary data for the first time everyone was a learner as they revealed their story.

Having considered the options, the author decided on a staged assessment, which proved surprisingly successful in engaging students, raising their aspirations, assessing their learning and providing useful information for Coventry University.

13.3 The teaching structure

The issue chosen for the survey was of topical interest to the university, being the impact on students of the transition to WebCT Vista, a Virtual Learning Environment, from WebCT Campus. The changeover had occurred that autumn, so the students had experienced WebCT Campus for a year and had themselves been faced with the change, which ensured that they were familiar with the scenario.

Students were enabled to carry out some of their assignment in class time. In order to give helpful feedback and ensure a speedy progression with the task, the assignment was tackled in five stages throughout the module, with different parts having different hand-in dates, and no extensions being allowed for the first two.

The five stages of the assessment were:

Stage 1:	Writing the background and objectives Handed in during Week 3	5 marks
Stage 2:	The Questionnaire Handed in during Week 4	5 marks
Stage 3:	Writing a proposed Sampling Strategy Handed in during Week 5	5 marks
Stage 4:	Data collection and entry This was carried out in Weeks 3 and 4 of the second term during the second teaching block. Failure to contribute to this part of the assignment deprived students of any further marks.	0 marks
Stage 5:	Report writing and critical evaluation Handed in at the end of the second term.	25 marks

The module began by discussing the difference between surveys and experiments and the importance of defining aims and objectives. As an exercise, students had to come up with aims and objectives for the Salk vaccine trail of 1954. This was one of the largest clinical trials undertaken showing 'that Jonas Salk's killed virus preparation was 80–90% effective in preventing paralytic poliomyelitis'. (Meldrum, 1998: 1223). The author has used this example on numerous occasions because of its historical interest, the sheer size of the samples and the ethical issues involved, which are usually totally disregarded by students, who are asked to consider possible methodologies before learning what actually happened. The methodology was then revealed and discussed, and a basic table summarising the outcomes presented, which students had to interpret and decide on the evidence whether the vaccine could be considered effective.

In the second week, the focus was on the students' own survey and the friendly client, the Deputy Director of the e-Learning unit, came to tell the students what he was particularly interested in learning about from a student perspective, which included the ease of navigation, any problems with access from home computers, which of the new facilities students had actually used and the uptake of the various help facilities, including the 24-hour hotline. Following this session, within a week with no extensions, students had to define the aims and objectives of their survey, and write some background to the issue of the transition from WebCT Campus to Vista.

In the third week, the principles of questionnaire design were studied, with many examples of professional questionnaires circulated and discussed. Students had to produce their own questionnaire for assessment, submitted electronically within six days, again with no extensions allowed.

Immediately following receipt of these questionnaires, the author produced a fairly long amalgamated version, from which a joint final version was created in class the next day. Groups of students discussed the questions displayed and, as a class, made decisions as to which were unsuitable and which needed rewriting. Using a wireless keyboard, the students themselves undertook the amendments; the final, agreed version was employed by everyone. The use of this wireless mouse and keyboard transformed the class into an active community of learners and achievers. A Wiki might also have made the compilation easier.

This final week of the first 11 teaching weeks was also used for discussing different sampling methods including simple random sampling, systematic sampling, cluster sampling, stratified sampling, quota sampling and fortuitous sampling. The practicality of data collection using each of these was experienced by students by means of a map of a hypothetical small town with a list of inhabitants and their views on taxation. Small groups of students adopted different sampling methods, carried out the 'survey' on the map, and compared the results, including those obtained by adopting the same method, for example, using a simple random sample but starting at different points in the random number table. Students also looked at the sampling strategies used by the Office of National Statistics surveys, online, and their problems with non-response.

Each student now had to write and submit a proposed sampling strategy for the class survey, together with a proposed timetable, and submit it for marking. More time was allowed for this, helped by the intervention of the Christmas break. Following this, a joint sampling strategy was adopted by the class, based on their individual ideas, which had been summarised by the author, who gave a firm steer during discussions away from the most time-consuming ideas.

Inevitably, many students thought that a random sample of students was the most appropriate sampling strategy. The university is divided into four academic Faculties, of varying sizes. The practicalities of obtaining a list of the target population ruled out a simple random sample, but the possibility of stratifying by size of Faculty were discussed. By actually carrying out the data collection themselves, the students were later able to reflect on the strategy chosen and on those deemed impractical.

Finally, it was agreed that for simplicity a quota sample would be adopted with the constraint that 50% of those asked should be female, 10% mature students and 10% overseas students, though these percentages could overlap so that one mature female overseas student would help satisfy all the specific criteria.

In two of the early weeks of the second term, everyone, including the author, collected the data and entered it into a prepared Excel spreadsheet. Pairs of students collected 30 responses each from specified spots around the university at specified times; the idea being that the author could do spot checks to verify the data collection method. The collection sites had been identified by the students in class, and the times of day and days of the week were also decided by them. To achieve an approximately equal number of responses from each of the different Faculties, the sampling was done outside the main entrances to each, and to the library. The author had merely to produce a list of times and places against which students signed themselves as pairs and collected from there at the specified time. The author collected data from 30 staff, which were analysed separately.

The first three stages of assessment proved extremely busy for the author, who was grateful for only 40 students, a few of whom dropped out at this point. Tight deadlines for work to be handed in also demanded tight return times – there were only 24 hours between receipt of the questionnaires and the production of the final version, though the marking was done later. The advantage was that everything was fresh in the minds of the students and they learnt by looking at the questions others had posed and the variety of formats offered. The lecturer was very happy to let the students take control, as a group, of the final questionnaire construction.

It is acknowledged that two hours to cover the theory of sampling is too short a time to do such a complex subject anything like justice, but by putting at least some of the theory into practice students can gain a greater insight into some of the problems.

13.4 The analysis

Students entered their own data into the prepared spreadsheet, all following the same written instructions, and submitted their spreadsheet electronically to the author, who amalgamated them all and then surveyed the mess. Students were allowed to view this initial file but were given a partly cleaned version in the next lecture and shown how to clean up the rest. It was a salutary experience for many, and a welcome to the real world of data collection. As one student commented, 'the biggest problem occurred in the data entry process, some students did not follow the coding sheet exactly or interpreted it in another way'.

The author produced a final cleaned version of responses from approximately 700 students, which was then used by the whole class for analysis. Initially this was carried out in Excel using PivotTables, which provided an excellent revision of this facility, but subsequently the analysis was completed in SPSS, being used as a vehicle to teach that package. There was also the opportunity to revise the principles of good charts.

Again, the use of a wireless keyboard and mouse was invaluable, enabling the lecturer to concentrate on the teaching and encouraging of students to explore SPSS as a class, while the students themselves did the mechanics of using SPSS. The keyboard and mouse were passed around pairs of students as the lecturer suggested different approaches, the students clicked and the whole class looked at the output and suggested interpretations. Separating the teaching from the management of the technology was a great release for the author, allowing far more attention to be paid to the students and their ideas and suggestions. By attending and making notes, students were able to write the outlines of their final report before they got to examine the data themselves, and had some idea of the structure of SPSS. These two whole class lectures on the use of SPSS, of approximately an hour each, were immediately followed by a workshop for another hour in which each student had access to a PC, with SPSS. In that hour, they reproduced the work covered in the lecture and added further to the analysis. Nothing was submitted at this stage.

13.5 The report and critical evaluation

A final session on report writing concentrating on the importance of structure, signposting and appropriate language for the intended audience, prepared students for writing their individual reports, which were submitted at the end of the second term.

The report was worth 20 of the 40 marks for the complete assessment. Five marks were allocated for each of the following:

- A coherent overall picture of the survey as a whole, with an introduction and conclusion and a clear structure to the analysis.
- Use of clear quantitative statements; appropriate summaries of the data.
- Appropriate investigation of some variables by factors.
- Well formatted tables and charts as appropriate.

An important part of assessment was the individual critical evaluation of the small-scale survey, submitted with the final report and worth five marks. It forced students to reflect on their experiences, from the design of the questionnaire and how much use it was in practice, to the problems of collecting data and entering them accurately. The comments showed that some students had found this useful as opposed to tedious, making various constructive suggestions relating to the questions used, the sampling strategy and the instructions for the inputting of data. Comments included: ‘Some data is not as useful as first expected. For example the percentage of people using certain features is not as helpful as how good or bad the staff and students find them’.; ‘I’ve had an insight into how members of the group can affect the deadline by not carrying out tasks correctly’.; and ‘The question “How many of your modules make active use of WebCT?” also appears to be slightly ambiguous. Nowhere is "actively used" defined and students may interpret this in different ways'.

Several of the reports were outstanding in their clarity and content and were greatly appreciated by Coventry University's e-Learning Unit. Because of this several students combined with the module leader to produce a poster summarising the headline results, Figure 13.1, for the university's Learning and Teaching conference that year.

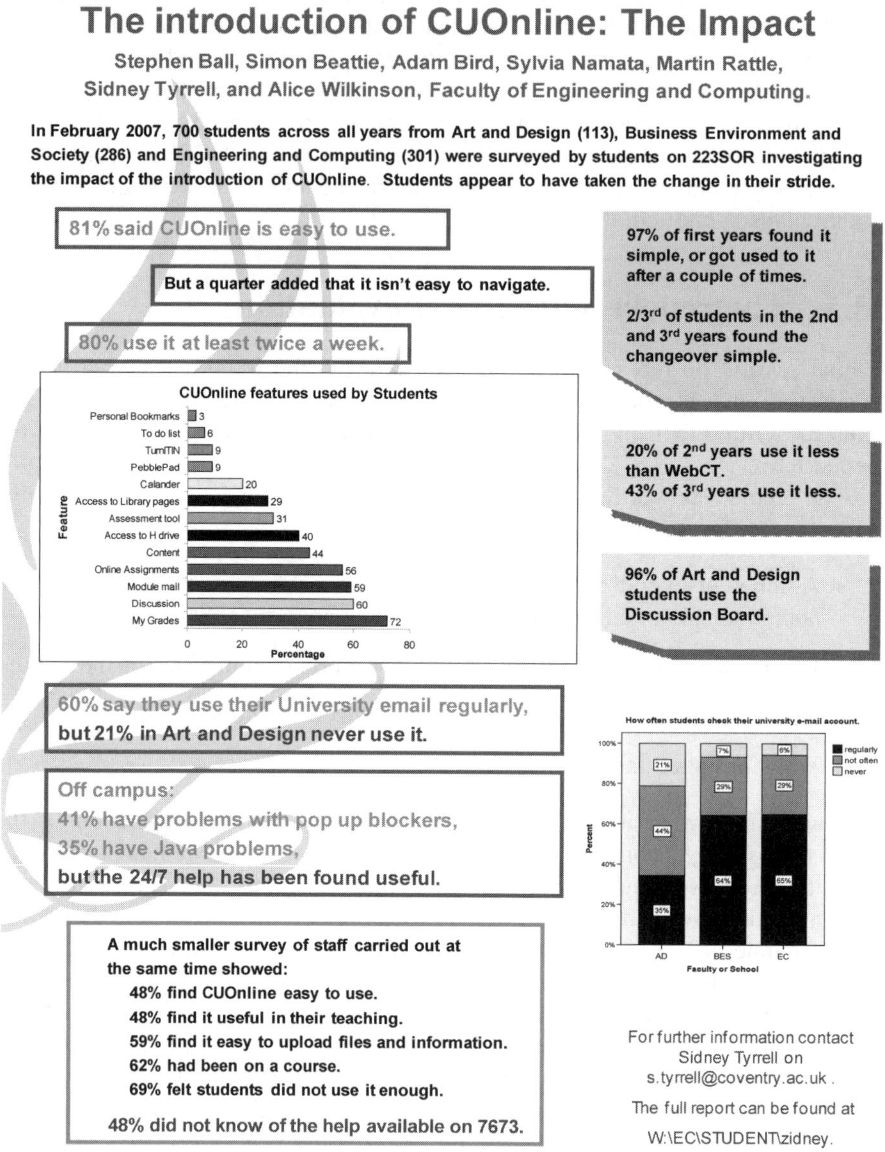

Figure 13.1 Poster summarising the impact of introducing CUOnline.

13.6 Conclusion

The staged assessment was successful in assessing the intended learning outcome of 'Design, conduct, analyse, report and critically evaluate a small-scale sample survey'. In terms of achievement the mean student mark for the survey part of the module was raised from 50% in a previous year to 57%, with more active student engagement and attendance.

By staging the assessment both students and author were kept on task, and by having a relevant topic and a genuine client students appeared more motivated to produce a report of professional standards. The use of a wireless mouse and keyboard enabled the class to become an active community of learners whilst preparing the final questionnaire and when using SPSS for the initial exploratory analysis.

The module was also successful in terms of the author's own experience in watching the class develop as a community of learners, free from handling the technology. Also, observing the difficulties students have in finding their way round a mass of data and gaining confidence in asking questions of SPSS using the wireless mouse has alerted the author to difficulties which are shared by many other students in the task of data analysis and suggested potential pathways to adopt. It has inspired the author to hand over the technology to students elsewhere, and the practice of giving timely feedback at a point when it can influence the next stage of the assessment has been adapted for a first-year skills module.

The final word, however, must be left to the students involved:

- 'Coursework was done in stages – very good! Less Stress!'
- 'The workload is not as demanding as other modules, BUT still a challenge! Like the use of technology, i.e. wireless mouse and keyboard, enjoy that part of the lesson'.
- 'Interesting subject material; real-world coursework'.
- 'While carrying out the survey I have learned how to carry out a live project and how to manage time to ensure milestones are met on time. The skill I believe I have learned and will be able to take away into the real world is how to interpret data into meaningful information'.

14

Evaluation of design and variability concepts among students of agriculture

María Virginia López, María del Carmen Fabrizio and María Cristina Plencovich

14.1 Introduction

Over many years of teaching statistics in a non-mathematical undergraduate programme of agricultural and environmental sciences, the authors have attained different learning results, using various teaching approaches. However, in spite of some learning achievements, it is found that even those students who get good grades in statistics courses seem to forget basic concepts learnt in initial courses.

One of the major difficulties that students have is the lack of understanding of variability, a difficulty also found by some other authors (Reading and Shaughnessy, 2004; Shaughnessy *et al.*, 1999). Students can easily use a method of multiple comparisons of means; but it is hard for them to understand that the sample mean has a known probability distribution that enables inferences to be made even from a unique experiment provided it is correctly randomised. delMas *et al.* (1999) and Saldanha and Thompson (2001) refer to the difficulties of understanding sampling distributions. This misconception hampers the distinction between estimator and parameter and underestimates the need for correct randomisation. This is not a minor problem, since the students will have to carry out experiments or take samples in their future professional practice. These analyses, no matter how complex they might be, cannot amend defective data.

Assessment Methods in Statistical Education Edited by Penelope Bidgood, Neville Hunt and Flavia Jolliffe

Mackisack (1994) describes a situation in which the background information and the description of the designs are written by the students. Her students were mathematics majors, some of whom might become practicing statisticians. Their background was therefore different from that of the students described here and they had different interests. On the other hand, according to Mead (1990), although it is possible to devise real experimental situations, the process of experimentation is expensive in both time and money. For this reason he used simulated situations for experimentation. He found that students do not have difficulty in imagining the practical details, and that they become very much involved in finding the correct solution.

14.2 Course description

Agricultural Engineering or Environmental Sciences programmes at the University of Buenos Aires have two compulsory statistics courses: General Statistics and Statistical Models. The former, with 80 hours of class contact time, covers descriptive statistics, probability and inferential statistics, including comparison of two means, simple linear regression and contingency table analysis. The latter, a 48-hour course, mainly focuses on the design of agronomic experiments, and students learn basic concepts of one- and two-way analysis of variance and multiple regression. Both courses are organised around lectures and practical classes. Lecture classes are large (60 to 120 students), whereas practical classes have smaller groups (up to 35 students). Some authors have referred to the lack of interest in quantitative subjects in agricultural university schools (Blanco *et al.*, 2006). However, students here are motivated and eager to learn new themes, because they are exposed to real agricultural situations in class.

In one of the first Statistical Models classes, students shown a field experiment conducted by a faculty research group have to identify concepts taught in lecture classes (such as correct experimental design), and recognise treatments and response variables. For example, one year students visited two rapeseed experimental sites, which, although related to the same crop, had different objectives and experimental units.

14.3 Course assessments

In the first year, students take General Statistics, a compulsory course which often has very large groups. The performance of students is assessed through a system of continuous evaluation over the entire period of studies, through midterm tests and assignments that are submitted in every class. If students gain 70% or above, they pass the course; below 40%, they fail. Students with intermediate performances have to sit for a final, integrated examination. Midterm tests have four or five problem-solving exercises. The final examination is generally multiple-choice.

There is a general consensus that some typical forms of assessment, such as multiple-choice examinations, are too narrow to provide sufficient information about student learning. However, multiple-choice tests may be used to capture students' reasoning and measure conceptual understanding (Garfield and Gal, 1999). In this course, application of statistical concepts is assessed by the midterm tests.

In the second year, students take the Statistical Models course, which involves various assessments. One take-home assignment consists of submitting a report with the results of analysing a real problem based on given data, using statistical software. This assignment is a good tool for assessing learning and justifies the use of open-ended items that require more time and effort in scoring them. Its main purpose is to expose students to real situations in order to determine if they are able to apply statistical concepts to solve a problem and communicate their results appropriately.

14.4 Methodology

The authors have found that students lack knowledge in applying concepts of randomisation and the inherent variability in sample statistics to real problems; hence they decided to carry out a teaching experiment. The objective was to carry out a learning activity in which students had to take their own sample instead of solving a written problem assignment with given data.

An assessment instrument was developed which evaluated competencies associated to students' future profession, including variability and randomisation concepts. As the situations depicted in the assignment were close to real professional practices, it was believed that they would be academically challenging for students, particularly compared to routine problem-solving exercises.

Students who participated in this study were attending the Statistical Models course in the first semester of their second year. There were about 290 students distributed between three lecture classes (designated A, B, and C) and eight practical classes (A1, A2, A3, B1, B2, C1, C2, and C3). All students had completed General Statistics and most of them were attending second-year courses on the Agricultural Engineering and Environmental Sciences programmes.

A three-stage experiment was designed. At Stage I, students had to solve a take-home assignment, individually or in pairs. Four of the eight practical classes had a traditional home assignment (called E1) and the remainder (A1, B2, C1 and C2) received the experimental problem (called E2). Both assignments dealt with agricultural or environmental issues. Sections 14.7.1 and 14.7.2 give examples of two equivalent home assignments, one traditional and one experimental, denoted E1 and E2 respectively.

In the experimental take-home assignment, students were given a diagram of a possible allocation of treatments to experimental units. Each experimental unit was composed of subunits, the values of which were generated by simulation from normal distributions with the same variance. Students had to generate their own sample for each unit by averaging the values of the selected subunits and, from these, to carry out the requested analysis. One of the objectives of the problem was to differentiate experimental units from observation units. Thus, the probability that two students groups obtained the same result was very low. After finishing the assignments, students submitted electronic files so that teaching teams could check that reported results were consistent with their data.

At Stage II, at the end of the course, students were assessed by a short test (E3) about variability in sampling statistics. All of the classes who had received the experimental assignment E2 (and due to operative reasons, only two of the others) sat the test; they had to answer four questions dealing with the definition of the experimental unit and variability. An example test E3 and the results for both groups from a particular year are given below.

At Stage III, six months after the end of the course, students were given a further test (E4), with the purpose of checking retention of learning. Two questions were sent via email to 206 students from the eight classes. The questions were similar to the last two of E3, but referred to sample means instead of treatment effects.

Pearson's chi-square test was used to compare the percentages of correct answers by problem type (significance level, $\alpha = 0.05$) for E3, and Fisher's exact test for E4.

14.5 Example from E3 and results

Here results from tests E3 and E4 are compared for the two groups – those taking the traditional assignment E1 and those taking the experimental assignment E2.

Example of test E3

> A medium-sized company located at Roberts wanted to find out Landrace hog weight increase by three diets in 30 days. The company had a large homogeneous hog herd. Hogs were assigned at random to 15 pens, 20 animals each. The response variable was the average weight increase in kg of four animals in each pen.
>
> a. Which is the experimental unit?
>
> b. Given that df Error = t (r – 1), which is the df Error value? Justify your answer.
>
> c. Two researchers conducted independent trials in the pens. One of them reported $\hat{\tau} = 2$ kg and the other $\hat{\tau} = 4$ kg. Why were their values different?
>
> d. Can we infer from these data that Diet 1 produced a 3 kg increase? Justify your answer.

Table 14.1 shows the number and percentage of correct answers by question for each type of assignment, experimental or traditional; 130 students were evaluated.

Table 14.1 Number of correct answers per question for E3.

Assignment	Question (a)	Question (b)	Question (c)	Question (d)
Traditional (n = 43)	33 (77%)	22 (51%)	23 (53%)	13 (30%)
Experimental (n = 87)	74 (85%)	65 (75%)	61 (70%)	46 (53%)
Chi-square statistic	1.37	7.21	2.94	5.95
p-value	0.242	0.007	0.086	0.015

Although there was no significant difference between the groups in the definition of the experimental unit (Question a), the students that received the traditional problem had more difficulty in applying this concept in the determination of the number of repetitions for the calculation of the degrees of freedom of the experimental error (Question b). There was no significant difference between the groups for Question c. The group who solved the traditional problem had more difficulties in differentiating parameters from estimators.

E4 was answered by only 33 of the 206 students contacted. Of these, 22 had performed well in the course, three had failed, and the rest had attained an intermediate level. The question about variability was answered incorrectly by only three students (two of whom had solved the traditional problem). Question b was solved by seven of the 17 who had solved the traditional problem (41%), and eight of 16 who had received the experimental problem (50%). There were no statistically significant differences (p-value = 0.732).

14.6 Conclusion

The development of assessment alternatives in our statistics courses has been based on the principles that assessment should reflect the statistical content that is most important for students to learn, and enhance learning of statistics, while supporting good instructional practice (Garfield and Gal, 1999). The possibility that students may select their own samples and obtain different results through such selection may be useful in learning the concept of variability. On the other hand, the difference between experimental unit and subunit leading to the determination of the correct number of replications – so frequently confused, even by professionals – can be clarified if students have to draw a subsample from the unit and then carry out the corresponding analyses. These concepts have to be discussed in a later class, after assessing students' responses to tests, in order that they see the differences in the results and get some feedback about variation. Unfortunately, due to lack of time, this debate did not take place in this study. When students were retested after six months (Stage III), they might be expected to have forgotten some concepts. The low number of students who responded to the second test may have been due to lack of time or interest, or because the test was not compulsory. This is in line with the findings of delMas *et al.* (2006), who state that, even when innovative teaching techniques are used, expected results are not always achieved. This is really frustrating, given the teaching effort. It also indicates that it would be better for students to obtain basic statistical concepts at high school, since it is difficult to achieve statistical reasoning within an undergraduate course on environmental or agricultural sciences, especially for students trained in deterministic ideas. The authors take the view that it would be of great value if students, in courses taken after General Statistics and Statistical Models, were to apply the concepts learned on variability and experimental design.

14.7 Traditional home assignment (E1): Full text

A trial was carried out with the purpose of determining the response of red raspberry (Rubus idaeus L.) 'Autumn Bliss' to a vermicompost application intercropped with lupine under greenhouse conditions. Vermicompost (VC) is considered one of the best organic manures, having good physical, chemical and microbiological characteristics; its use prevents crust formation, improves soil structure, aeration, water retention and drainage, and increases ionic exchange and phosphorus availability. Lupine (Lupinus mutabilis Sweet) is a species from the Andes frequently used to intercrop with potato, corn, quinoa and other crops. From remote times, it has been used to improve fertility and fix atmospheric nitrogen; due to the depth of its roots, it can extract nutrients from deep soil layers.

In this trial, three levels of VC were studied (0, 30 and 90 gr/pot) and two association levels (without and with lupine). As vegetable material, cold-treated adventitious buds of red raspberry were used. Light and temperature conditions were similar in all the greenhouses. After 90 days, the phosphorus level was evaluated (P) in mg/g in the foliar tissue. Results are presented in Table 14.2:

Table 14.2 Trial results.

	VC Levels											
Lupine	0				30				90			
Without	1.41	1.51	1.48	1.48	1.66	1.53	1.62	1.59	1.90	1.78	1.89	1.83
With	1.51	1.57	1.62	1.58	1.81	1.70	1.87	1.94	2.16	2.24	2.11	2.30

a. Write the corresponding model and describe each component according to the problem. What are the model assumptions?

b. Which is the response variable? Which is the experimental unit?

c. Describe factor/s, levels and treatments.

d. Design a diagram of the greenhouse with treatment assignment.

e. Draw a graph to explore the presence of a different response to VC by plant intercropped or not intercropped with lupine. Make some comments about your findings.

f. Calculate the simple effects of intercropping with lupine for VC levels 0 and 90. Analyse the results.

g. Check your observations on the graph (e) against the corresponding hypotheses.

h. Write thorough conclusions.

This problem has been adapted from Jara-Peña *et al.* (2002).

14.7.1 Experimental Home Assignment (E2): Full text

Dry cassava leaves could replace soya beans and fish flour in concentrated food production, thus diminishing fowl feeding costs remarkably. Cassava leaf protein content is relatively high; hence it may replace more expensive protein sources. In an experimental station a trial about chicken feeding with concentrates was conducted. Concentrates included cassava flour in different proportions (0%, 10% and 30%). Researchers wanted to know if the effect of incorporating cassava leaves was similar for two of the commonly used breeds in the hatchery. Both were included in the trial. The data provided in Table 14.4 has 10 chicken weight increases (grams) for each pen, measured at the eighth week of trial onset. Each pen has chickens of the same breed and a feeder with the cassava percentage assigned. Select four chickens from each pen at random and calculate the average weight increase for the analysis. Table 14.3 shows the name given to each treatment and how they are indicated in the data table.

Table 14.3 Name of each treatment and their identification.

	Breed	
Cassava %	1	2
0	▲T11	■T12
10	•T21	♦T22
30	◄T31	*T32

a. Write the corresponding model and describe each component according to the problem. What are the model assumptions?

b. Which is the response variable? Which is the experimental unit?

c. Describe factor/s, levels and treatments.

d. Draw a table for your data.

e. Draw a graph to explore the presence of a different response to cassava percentage by breed. Make some comments about your findings.

f. Calculate the simple effects at 0% and 30% cassava in Breed 1. Analyse the results.

g. Check your observations on the graph (e) against the corresponding hypotheses.

h. Write thorough conclusions.

Table 14.4 Data sheet.

■1779.78	■2165.50	▲1905.33	▲1952.22	◀1008.93	◀1138.11	◀760.68	◀589.79
■1775.87	■2332.77	▲1916.12	▲1980.38	◀1238.40	◀1174.33	◀645.26	◀963.93
■1965.00	■2324.52	▲1692.58	▲1930.32	◀1429.15	◀1395.09	◀1194.07	◀877.21
■2248.23	■2326.62	▲2085.66	▲1725.45	◀1231.62	◀1526.25	◀702.20	◀1064.98
■2203.06	■2392.58	▲1800.01	▲1778.96	◀1003.35	◀1300.35	◀818.39	◀375.76
♦2345.20	♦2101.73	▲890.06	▲842.05	•1595.82	•1743.97	*2158.15	*1777.89
♦2208.74	♦2306.25	▲795.08	▲1070.81	•1895.33	•1958.91	*1962.14	*1828.27
♦1933.77	♦2716.35	▲756.85	▲798.58	•1843.20	•1976.96	*1937.03	*1881.54
♦2037.33	♦2305.30	▲775.92	▲901.38	•1361.52	•1845.81	*1861.60	*1947.63
♦2504.77	♦2504.81	▲1118.45	▲1206.58	•1741.31	•1923.57	*1654.67	*2055.25
◀1461.94	◀1038.20	■1936.66	■2016.84	•1093.94	•784.13	•2053.66	•1975.32
◀1725.33	◀1533.87	■2085.74	■2262.24	•951.07	•943.35	•2143.81	•2300.87
◀1294.05	◀1628.59	■1946.36	■2294.98	•722.49	•822.80	•1899.00	•1969.91
◀1316.04	◀1611.78	■2386.56	■1983.23	•875.15	•1096.42	•1883.39	•2093.77
◀1535.65	◀1011.03	■1854.83	■1968.61	•666.05	•608.67	•2060.26	•1943.62
*1378.89	*1589.03	♦1522.39	♦1705.61	♦1847.96	♦1961.78	◀1012.64	◀980.30
*1746.70	*1555.54	♦1439.36	♦1532.03	♦1895.09	♦1890.34	◀1049.94	◀1135.65
*1614.98	*1523.22	♦1636.67	♦1894.72	♦2202.44	♦2078.74	◀1075.27	◀1190.16
*1737.82	*1216.22	♦1344.47	♦1583.94	♦2180.14	♦2061.36	◀1115.35	◀1313.12
*1512.06	*1379.34	♦1617.27	♦1556.78	♦1699.29	♦1852.49	◀1079.15	◀1288.41
♦2531.23	♦3015.07	•1272.11	•1557.32	■3127.63	■2855.21	▲1831.49	▲1658.57
♦2494.60	♦2213.47	•1521.71	•1749.91	■2418.10	■2359.29	▲2176.47	▲1987.79
♦2368.45	♦2596.59	•1239.49	•1244.36	■2657.50	■2630.54	▲1613.44	▲2530.63
♦2802.96	♦2405.96	•1544.06	•1311.73	■2835.42	■2536.06	▲1769.08	▲1869.41
♦2533.03	♦2636.15	•1634.18	•1599.26	■2560.19	■2541.60	▲2254.13	▲1940.04
*1798.55	*1635.40	■2670.15	■2332.32	*1367.96	*1271.01	▲1360.51	▲1578.15
*2163.02	*1567.65	■2224.04	■2724.10	*1302.01	*1238.56	▲1356.92	▲1287.35
*2013.34	*1854.84	■2622.78	■2711.28	*1350.52	*1238.59	▲1630.89	▲1761.84
*1836.12	*2170.88	■2277.70	■2616.46	*1441.99	*1362.70	▲1720.99	▲1556.34
*1496.33	*1935.88	■2332.48	■2457.65	*1130.41	*1145.61	▲1509.10	▲1689.69

15

Encouraging peer learning in assessment instruments

Ailish Hannigan

15.1 Introduction

Peer learning, as defined by Boud *et al.* (1999: 413), refers to 'the use of teaching and learning strategies in which students learn with and from each other without the immediate intervention of a teacher'. This definition includes the concept of cooperative learning where students work together to maximise their own and each other's learning. It also includes the concept of reciprocal peer learning where students act as both teachers and learners. Examples of situations where peer learning can occur include group projects, study groups, student-led workshops and peer feedback sessions in class.

The skills associated with peer learning are worth encouraging, particularly for statisticians, and are often demanded by graduate employers. They include critical enquiry, reflection, communication skills, the ability to collaborate with others in a team environment and learning to learn. Communication, in particular, is considered fundamental in statistics; statisticians often need to identify a client's needs at the start of a project and then convey the results of the analysis to the client in non-technical terms (Begg, 1997). Statistical practice can be an inherently cooperative enterprise with statistician and client working together and statistics instruction should reflect that (Roseth *et al.*, 2008).

Boud *et al.* (2001) also suggest more pragmatic reasons for the popularity of peer learning in universities where lecturers are asked to teach increasing numbers of students with limited resources. Peer learning appears to maintain student learning without more input from staff. It also encourages graduates to

Assessment Methods in Statistical Education Edited by Penelope Bidgood, Neville Hunt and Flavia Jolliffe

become lifelong learners, which is essential at a time when technical knowledge is constantly changing. The nature of the student body has also changed, with students often being more confident and critical and expecting direct and active involvement in their education. Increasing numbers of students are returning to learning later in life and peer learning and support is vital for such students (Anderson and Boud, 1996).

Peer learning has also been shown to increase students' long-term ability to retain material in statistics courses (Kvam, 2000). Two classes of students participated in his study, both taking an undergraduate engineering statistics course with the same recommended textbook. One class was taught using traditional lecture-based learning; teaching in the other class emphasised group projects and peer learning. Students were tested immediately after the course ended and again eight months later to measure how much of the original material was retained. Retention of subject material was higher in the peer-learning group compared to the traditional lecture-based group, particularly for students with average or below average scores in the first test.

The potential effectiveness of peer learning is well documented but small-group work can give rise to negative experiences (Pauli *et al.*, 2008). Frequently raised concerns regarding group work include the potential for conflict between group members and the possibility of individual members not doing their share of the work. 'Freeloading' can occur when one or more students fail to contribute to the group effort because they assume the work will be done by more talented or more motivated group members. Care needs to be taken in the design and execution of group projects, otherwise the problems that may ensue can reinforce the 'myths' of group work, for example, that the composition of the group unfairly affects one group over another, or that unequal contributions from team members within a group unfairly affects grades (Livingstone and Lynch, 2000).

Assessing peer learning can also be problematic and can easily inhibit the processes it is designed to enhance if it is not implemented sensitively. The skills encouraged by peer learning, for example communication, critical analysis and group work are traditionally difficult to assess and staff assessment skills in these areas are less developed. It is possible to encourage peer learning without including it in a formal assessment but Boud *et al.* (1999) suggest that assessment of peer learning needs to be considered so that it is valued, commitment is recognised and important educational outcomes are addressed.

15.2 Study groups and assessment

Roseth *et al.* (2008) suggest several ways of encouraging peer learning in classes, including establishing a consistent base group. At the beginning of class, members of a base group can check in with each other, ask each other questions, compare notes and discuss their preparation for class. Base group members give themselves a daily rating on how well prepared they are for class. The base group then calculates the group's average preparation rating and charts it over time.

By monitoring the group average over time, students gain awareness of their study habits and also recognise that they are accountable to their peers.

The GIG procedure (Johnson and Johnson, 2002) proposes Group preparation, Individual assessment and Group assessment as a method of assessing how much each student knows and making each student accountable to peers for his/her performance on the assessment. Students meet in their study groups and are given test questions and time to prepare for the test. The students discuss each question and come to a consensus about the answer. The goal is to ensure all students can answer the questions correctly. Each student takes the test individually, making two copies of their answers – one for the instructor and one for the group discussion. Students meet again in their study groups and take the test as a group, ensuring that all members can explain the answer and understand the underlying rationale and procedure.

Roseth *et al.* (2008) suggest ways in which the GIG procedure can be used in practice; for example extra credit can be awarded if every member of the study group scores above a certain threshold on the individual test. The assessment can include an individual portion and a group portion. For the group portion, one randomly chosen member of each group will have their assessment graded and their score is given to the group as a whole. This structures one of the requirements of cooperative learning: knowing that your peers' success depends on your own.

Chance (1997) uses pairs of students working together in computer laboratory sessions to facilitate peer learning in introductory statistics courses. In the laboratory sessions, students work with computers directly, producing statistical output and interpreting that output. Each pair of students is required to produce a report on the computer laboratory, identifying, executing and justifying the relevant statistical procedure. Half of the marks allocated to the report are given for discussion of the results. Writing up the laboratory sessions teaches students to base their conclusions only on the statistical evidence and gives the students practice with the language of statistics. Working in pairs allows the students to debate ideas with each other and Chance suggests that this is when the most significant learning occurs. The laboratory report that receives the highest grade each week is made available to all students so that they can see the work of their peers.

15.3 Group projects and assessment

The assessment of group projects is challenging and involves assessing two separate but related issues: the nature and quality of the group process and the quality of the product of the work of the group (Jolliffe, 1997). Assessing the product is generally more straightforward than assessing the process. Often, a group mark is allocated to the product and each member of the group receives that same mark, which counts towards their final grade for the course. Alternatively the overall mark for a group can be distributed according to the contribution of each individual to the project. The first approach assumes that all group members contribute equally and can give rise to complaints from students that this is not the case. The second approach requires determining the contribution of each group member and often involves peer assessment.

A staged assessment approach with an individual and a group component can also be used, for example the 4P model, as described by Starkings (1997). Marks are allocated for each stage of the project: the Project log; Project report; Practical development; and Presentation. Each group member is required to keep a logbook over the lifetime of the project. This logbook is assessed to determine the individual's effort, commitment, quality of work, and integration and cooperation with the rest of the group. The rest of the project, for example the final report, is assessed as a group.

15.4 Peer assessment

Peer assessment involves students assessing the 'amount, level, value, worth, quality, or success of the products or outcomes of learning of peers of similar status' (Topping, 1998: 250) against a set of clear criteria or standards. It can range from students marking multiple-choice questions against a template, or providing editorial comments on a written task, through to judging the quality of an oral presentation (Bilgin and Fraser, 2007). Criticism of peer assessment includes concerns over reliability and validity, since students may not have the skills or understanding to make judgements about the work of others or provide useful feedback. Boud *et al.* (1999) suggest that peer assessment can inhibit cooperation and that students find a contradiction between a learning process of working together to help each other and an assessment process which can encourage competition.

Peer assessment in group projects requires group members to evaluate the contribution of other group members to the project. Often a group member is asked to assign anonymously a number out of 100 to the contribution of each of the rest of the group and the average score for each group member from their peers is used to determine their proportion of the mark allocated to the product. While this method lack openness, it may be easier for students compared to openly discussing and coming to a consensus about the contribution of each group member. A confidential peer assessment form can also be used, such as the one devised by Conway *et al.* (1993). They proposed that the final mark given to an individual is the product of an individual weighting factor (IWF) and the final group mark. The IWF is calculated using the ratings given by the rest of the group in the peer assessment form. Chance (1997) suggests splitting the grade for group projects into an individual grade (15%) and a group grade (85%). The individual grade varies across group members and is based on confidential input from the group at the end of the project. Consistent negative feedback from the group on a particular member results in a lower individual grade for that member. Contracts can also be used at the start of a group project to identify which parts of the project each person will work on and what their contribution will be. Group members who do not fulfil the terms of that contract are given a lower individual grade for the project.

15.5 Peer feedback and self-assessment

Peer feedback is a communication process through which learners enter into dialogues related to performance and standards. Liu and Carless (2006) define it as primarily about rich, detailed comments but without formal grades, and suggest it has greater potential for learning than peer assessment. Peer feedback processes can develop skills such as critical reflection, listening to and acting on feedback, sensitively assessing and providing feedback on the work of others. Peer feedback can also enable students to better self-assess themselves, as some skills are common to both peer and self-assessment. Self-assessment is a key skill for lifelong learning, since learners need to be realistic judges of their own performance and monitor their own learning. Peers may provide rich information, which can then be used by individuals to make their own self-assessments (Boud, 1995) and follow up with actions to improve their work.

Sisto (2009) used peer feedback of project presentations to help students develop critical listening skills as consumers of statistical information. Students had to give two positive comments and two constructive comments on how the project could be improved for each project presentation. Each student received marks on the basis of their critiques and this was used to adjust the overall group project grade for the student. Each project group also received an anonymous summary of peer comments. Sisto suggests that students are more open to feedback from their peers than their teachers and by doing the peer assessments, students are developing transferable skills in evaluation and critical listening.

15.6 Encouraging peer learning in an assessment instrument

A survey-based group project is used by the author to encourage peer learning in an assessment instrument. This assessment includes many dimensions: students select a sample of fellow students to survey; design a questionnaire; implement the survey; analyse the data collected using a suitable software package; and write a report and present the project to the class. The written report includes criticism of the project by the group, in the form of a section on 'ways we could have done this project better'. The oral presentation of the project to the class includes receiving peer feedback. The assessment aims to cover the key elements of assessing student learning of statistics – factual knowledge, understanding of concepts, computational ability, appropriate application of techniques, and the practical skills of doing and communicating statistics (Jolliffe, 1997).

The group dimension to this project, the requirement for self-criticism by the group and peer feedback from the class encourages peer learning. It provides an opportunity for students to interact outside the classroom, discussing what questions will be included in their questionnaire, how they will implement the survey and how to present their work. Each group is implementing a survey to answer the same research question, and so the oral presentations allow other groups to compare different approaches taken to the same problem. At the end of the oral presentation by a group, fellow classmates ask questions and make constructive comments about the project, which gives the group the opportunity to learn from peer feedback. The oral presentations are also used to evaluate the group process and ask individual group members about their project to determine their individual contribution.

The group size is usually three or four (depending on class size). Smaller groups are considered to be more successful than larger ones – in general, as the size of the group increases, so too do the resources required to help the group succeed (Johnson and Johnson, 2006). To date this assessment instrument has been used for first-year mathematics degree students and graduate students from non-mathematical backgrounds, both taking an introductory course in statistics. The graduate students are of different nationalities, academic backgrounds and with varying amounts of work experience. These students are generally randomly assigned to their project groups to give them the opportunity of working with students they do not know or who come from a different background. The quality of the group process is generally better in the more mature graduate students compared to the first-year group. Gatfield (1999) reported significant differences in the level of satisfaction with group projects and peer assessment of students having work experience compared to those without.

In the first-year class, students are given a two-week period at the beginning of term to form their own project groups. After that period, any students who do not belong to a project group are assigned to existing project groups. The unassigned students tend to be weaker students with poor attendance. If left to their own devices, these weaker students would struggle in a group of their own; assigning them to existing groups enhances their peer learning opportunities, although it can lead to negative group experiences, such as freeloading. Negative group experiences were more prevalent in the first-year class, as were issues such as poor attendance or dropout. Some groups reported never or rarely having met certain members of their group.

Part of the project involves students analysing the data collected from the questionnaires using a suitable statistical software package. Students are allowed to use scheduled computer laboratory sessions for the course to analyse the data and ask questions about the analysis. Also, each group meets with the lecturer, outside of scheduled class time, before the project is submitted. Monitoring and, if necessary, intervening in each group is considered a key step in cooperative education (Johnson *et al.*, 1998). The computer laboratory sessions and group meeting give evidence of how well the group is working together and also the individual contributions of group members. It also provides an opportunity to help groups that are not working well to improve their teamwork skills. This can include encouraging discussion and joint decisions on, for example, how to select the sample instead of letting one stronger student make the decision for the group. It can also include identifying students who are not contributing to the group, contacting them on behalf on the group, finding out why they have not contributed and suggesting, with the cooperation of the group, ways in which they might contribute before the project is submitted.

Self-criticism by the group and peer feedback are important parts of the assessment. Students included sections in their project report on 'ways we could have done this project better' or 'what we have learned'. This gave evidence of groups which had worked well, for example: 'This project, for this particular team, brought together four different viewpoints, four different ideas, four different critical analyses and amalgamated them into the project you find before you. We are pleased with what we have produced'; and from a group where the process was not as successful: 'Unfortunately we came across a problem ... most likely due to a communications error between project members'. It also gave evidence of self-criticism of their survey, such as: 'We now realise that ... was not the best place to collect random samples', 'We mistakenly underestimated the importance of questionnaire design. A much larger percentage of our total time ... should have been spent on this', and 'If we were to design the questionnaire again we would put more emphasis on clarity'.

Students were also given the opportunity to give feedback to other groups after the oral presentations. The graduate students were more likely to offer feedback to other groups and also seemed to be more open to learning from feedback offered to them. First year students seemed to benefit most from observing how their peers had tackled the same problem in the oral presentations but were often reluctant to offer peer feedback.

A variety of different ways have been used to assess the project over the years. Initially, only the product – the final report – was assessed, and all members of the group were given the same mark, which counted towards their final grade for the course. The author has also experimented with peer assessment using the group's anonymous assessment of the individual contributions of each group member. First-year students were particularly reluctant to award less than the total to any of the group members, except where that group member had rarely or never turned up to group meetings.

Currently, a version of the staged assessment method is implemented, with marks allocated to two stages of the project: some on a group basis, for example, the report submitted; and some on an individual basis, for example, performance in the oral presentations and scheduled laboratory sessions. The first stage involves designing and piloting the questionnaire, deciding on a suitable method of sampling and giving the final questionnaire to a sample of 40 students. A report is submitted after the first stage, which provides the opportunity of monitoring the output from the group and meeting with groups which have not performed well, before the next stage of the project is submitted. The second stage involves analysing the data collected using a suitable statistical software package, writing a report on the results and presenting the project to the rest of the class. The oral presentations and computer laboratory sessions give the opportunity to see the evidence of peer learning, how well the group process is working and also the individual contributions to the project.

While the project is a more time-consuming form of assessment than a traditional examination-based assessment, the benefits for the students and teacher are obvious. The students get the opportunity to put statistics into practice with the help of their peers and have also allowed their data, questionnaires and sampling methods to be used as a rich bank of examples for other students. Whilst a group project does not necessarily include peer learning, designing the project to encourage it has been worthwhile.

15.7 Conclusion

The skills associated with peer learning are particularly important for statisticians and statistics instruction should reflect the cooperative nature of statistical practice (Roseth *et al.*, 2008). Assessment instruments, including group projects and tests given to study groups, can be designed to encourage peer learning. Whilst peer learning does not involve the immediate intervention of a teacher, it is important that the teacher continually monitors and intervenes if necessary in each group to maximise the students' learning. A variety of different ways exist to assess group work, including group assessment, peer assessment, peer feedback and self-assessment. Which method works best in any given course may depend on the background of the students and experience of the teacher. Given the changing nature of the student body and the need for them to become lifelong learners, encouraging peer learning is worthwhile. Assessment is the single most powerful influence on learning in formal courses (Boud *et al.*, 1999), so designing assessment instruments to encourage peer learning is essential if the skills associated with it are to be developed.

16

Inquiry-based assessment of statistical methods in psychology

Richard Rowe, Pam McKinney and Jamie Wood

16.1 Background

Psychology is a young science and there is great potential for researchers to advance the frontiers of knowledge through quantitative research. Nevertheless, psychology students are often surprised by the centrality of research methods and statistics in their undergraduate training. Their backgrounds are heterogeneous; although a few have previously studied advanced mathematics and many have previously studied psychology, a substantial proportion comes from an arts or humanities background.

16.1.1 Statistics in psychology

It is common for students to have difficulty in engaging with statistics lectures. Typical comments from student evaluations include:

- 'The module was understandably tedious in places'.
- 'Although very boring, this module has helped me grasp statistical tests'.

Although lecture courses are often titled 'Research Methods' or 'Experimental Design and Analysis', it is not uncommon for their focus to be on statistical

Assessment Methods in Statistical Education Edited by Penelope Bidgood, Neville Hunt and Flavia Jolliffe

analysis rather than experimental design. Usually, traditional didactic lectures are used to explain statistical tests, and computer classes are often attached to these lectures to allow application of techniques covered using appropriate software. Assessment is frequently by examination, with questions commonly presenting a fictional experiment with some associated computer output. Students are graded on how well they can interpret the output. Despite a lack of enthusiasm for statistics, students often have sufficient study skills to perform well in examinations. However, their knowledge may not generalise well to new problems, for example, in managing the design and analysis components of their self-directed empirical work.

This situation provided an opportunity to revise the teaching and assessment of statistics. This chapter reports an inquiry-based learning (IBL) approach to teaching and assessing statistical methods on a first-year course with approximately 100 students. In particular the aim was to engage students in the research process, demonstrate its value in advancing knowledge and produce graduates who are more independent researchers. This endeavour was guided by the framework of IBL.

16.1.2 Inquiry-based learning and statistics in psychology

IBL may be defined as learning involving a process of self-directed exploration. Rather than passively receiving information through didactic methods, students are provided with open-ended scenarios where different approaches may lead to equally valid solutions and students have the freedom to choose the methods employed (Kahn and O'Rourke, 2005). Fisher and Moore (2005) report that IBL has been used to apply psychological theory to practice at the University of Plymouth, United Kingdom. In the study concerned, as well as linking theory and practice effectively, the IBL process facilitated the development of a range of graduate skills, for example, improved problem-solving skills and confidence. In a greater number of cases, problem-based learning (PBL), which is closely linked to IBL, has been usefully employed as a method of engaging students with disciplinary content, skills and methodologies on psychology courses (Willis, 2002; Pond III, 2004). Particular emphasis has been placed on the use of authentic PBL scenarios and tasks for the development of practitioner and professional competencies, including student ability to direct their own learning, especially in the field of clinical psychology (Huey, 2001; Dahlgren and Dahlgren, 2002; Reynolds, 1997).

In some senses, IBL may be thought unsuitable to teaching elementary experimental analysis, as many statistical questions will have a single correct answer in terms of test choice, method of application and interpretation. However, the broader research process is inherently inquiry-based and choice of research question and method of approach are both open-ended activities that determine the appropriate analysis. It is as a component of the entire research process that academic psychologists apply statistics. Through employing IBL, first-year students were provided with a flavour of this context in their research methods training.

In order to develop independent research skills, this project aimed to provide the students with tasks similar to those that academics would need to undertake in order to conduct research. The principles of aligned teaching emphasise that deep learning is more likely to occur in situations where the curriculum, teaching methods, assessment procedures, context of tutor-student interactions and the institutional climate are aligned with each other (Biggs, 2003). Brew and Boud (1995: 70) argue that, 'Doing research demands a deep approach to learning. Researchers therefore model, in their own work, learning approaches which it is desirable for students to emulate'. Linking teaching and research is discussed at length in Jenkins *et al.* (2003) and it is concluded that engaging undergraduate students in the research culture of the department is beneficial on multiple levels.

Psychological research is almost never conducted in isolation. Single-author papers are very rare in the quantitative psychological literature. Therefore, the IBL activity was designed to be strongly collaborative at all stages, including assessment. Sander *et al.* (2000) have shown that students expect to be taught via formal lectures at university but prefer to learn via group-based activities. Collaborative inquiry involves students working together to approach a task or question, generate discussion based on their experiences and reading, and negotiating through the created shared knowledge towards a joint approach to the problem. Constructivist theories of education propose that an environment that fosters deep approaches to learning can be created through the use of peer collaboration, as the dialogue it entails can shape, elaborate and deepen understanding according to Biggs (2003). Collaborative techniques have been widely used for statistics teaching and this has been found to reduce students' anxiety and improve abilities to build statistical skills and knowledge (Delucchi, 2007). Beyond improving statistical and research methods skills, collaborative inquiry also helps students develop cooperative team working skills that are required for most careers (Race, 1999; Biggs, 2003).

16.1.3 Aims of the project

This project formed one strand of a larger departmental project to build on existing excellence in IBL entitled PEBBLE (Psychological Enquiry-Based Learning). The PEBBLE project was funded by the Centre for Inquiry-based Learning in the Arts and Social Sciences (CILASS), a Centre for Excellence in Teaching and Learning (CETL) based at the University of Sheffield, United Kingdom, supported by the Higher Education Funding Council for England (HEFCE). Project funds were used to buy staff time for the curriculum design activities and capital funds were used to purchase ten laptop computers to be used in delivering the project.

The project introduced an experimental design and statistical analysis activity to the first year tutorial programme. This was integrated with the associated research methods lecture course. Lectures addressed descriptive statistics, experimental design, t-tests, Pearson correlation and simple contingency table analysis. The project was designed to introduce students to the whole research process, including selecting a research question to address, formulating a hypothesis, designing an experiment, choosing a statistical analysis, running the analysis, and reporting and interpreting the results. The rationale was that when students could see statistics embedded in the whole process then they would be more able to generalise their statistical skills to novel research situations in the future.

16.2 Methods

In this section we discuss methods of assessment and the assessment materials we use.

16.2.1 Initial tutorial

On the course there are approximately 100 students who are divided into tutorial groups of four to five, led by postgraduate tutors under the supervision of the first-year research methods lecturer. More than 20 postgraduate tutors (some running two groups) studying for both taught and research-based higher degrees in psychology are employed to lead these tutorials. Some were taking an MSc Research Methods in Psychology course that included a postgraduate tutor training module, and these keep a reflective diary regarding their experiences of small group teaching as a course requirement. All tutors attended a one-hour training session with the research methods lecturer to introduce them to the tutorial activities and ensure a standard approach.

The design of the IBL tutorial programme for the statistical methods module gave postgraduate tutors an opportunity to link their teaching with their research, a feature of IBL that, it has been argued, has positive benefits for both tutors and students (Brew, 2006). Prior to the tutorial, the tutors were asked to prepare three questions from an area of research with which they were familiar. These were submitted to the research methods lecturer for screening and either accepted or returned for revision. The questions were presented to the students at the start of the tutorial and they were asked to choose to focus on one of the questions. The tutor led a group discussion of the issues involved in designing an experiment to address the chosen research question. Topics covered included hypothesis formation, the advantages and disadvantages of within- and between- participant designs, choice of dependent variable and potential levels of the independent variable. As the discussion progressed, the group filled out a generic research proposal form that contained all the information necessary for data to be simulated for their design.

16.2.2 Data simulation

On the basis of the submitted design form a data set was generated for each tutorial group. The 'drawnorm' command of Stata (StataCorp, 2003) was used for data simulation. Data were generated to have means and standard deviations that were appropriate for the measures chosen. These were based on the knowledge of the tutors and the research methods lecturer. In many cases more variables were created than specified in the original design sheet, to allow the full range of statistical tests covered in the first-year lecture programme to be applied to different aspects of the data set. A mix of significant and non-significant relationships was specified in each data set.

16.2.3 Assessment materials

The assignment instruction sheet gave a description of the variables in the simulated data set. Following this were five questions that the students needed to answer using the data set. The first four required one each of a correlation, paired t-test, independent samples t-test and contingency table analysis to be answered correctly. The fifth question asked the group to 'Choose one further analysis to run based on your data set and write up the results'. This would involve repeating one of the tests already used in the assignment, as all the tests the students had been taught had already been covered. Students were told to write up the answers to all five questions using Word and include graphs and tables of descriptive statistics as appropriate. They were also instructed to quote statistical test results in the format of the American Psychological Association and provide brief interpretation.

The students had been informed that notes and textbooks could be consulted and this was confirmed on the guidance sheet. It was also stated that the tutor would provide assistance with running analyses.

16.2.4 Example assessment

In an example tutorial group, the students designed a study to examine the relationship between driving aggression, as measured by a standard questionnaire available in the literature, and age, using correlation. In order to allow questions requiring the full range of statistical methods covered in the course, the data set was expanded. The variables provided included a normally distributed driving aggression score before and after a driver attitudes training programme, a binary variable indexing whether the driver had ever had a crash, driver gender and age. The four specific questions and associated analyses are shown in Table 16.1. This data set offered a range of possible questions that students could address for the analysis of their choice. These included comparison of crash-involved and non-crash-involved drivers in driving aggression and gender differences in post-intervention driving aggression. The group chose to test whether there was a significant difference in age between crash-involved and non-involved drivers, using an independent samples t-test.

Table 16.1 Questions set in an example assessment.

Question	Statistical test
Is age related to pre-intervention driving aggression?	Pearson correlation
Are male drivers more aggressive than female drivers at pre-intervention?	Independent samples t-test
Are male drivers more likely to have been involved in an accident?	Contingency table analysis
Did the anger management programme reduce driver aggression?	Paired samples t-test

16.2.5 Second tutorial

Each tutorial group was provided with two laptop computers running SPSS and Word. Memory sticks were available to facilitate data transfer between computers. The simulated data set was preloaded onto both computers. Tutors were instructed to ensure students had 50 minutes to work on the project, allowing ten minutes per question. Room bookings were for one hour, so ten minutes was allowed for changeover. Tutors instructed their groups that they could use their resources as they chose; they could all work on each question together or they could split into two groups and apportion different questions to each group. At the end of the session the students saved their completed Word document and this formed their submission for the assignment. There was only one submission from each group and all students received the same mark.

16.2.6 Marking

The first four questions were marked on whether the correct test had been chosen to answer the question, whether it had been conducted properly, reported correctly, supported with appropriate graphs/tables/descriptive statistics, and interpreted accurately. The question asking the students to choose their own test was additionally assessed on whether their choice of question was appropriate.

The collaborative aspect of the assessment had implications for the marking strategy. The students worked in groups of four to five, and so there were fewer scripts to mark than in a traditional examination. Individual comments are not usually provided on examination performance but they were given for this assessment, as it was also being treated as a teaching opportunity. The lower volume of scripts reduced the time commitment required to provide detailed comments. Marking was conducted using an Excel spreadsheet where various criteria for each question were identified as fulfilled or not. The spreadsheet combined these scores and translated them into an assignment mark. Each cell was also linked to a cell containing a comment regarding that criterion, with the returned comment differing depending on whether the criterion was fulfilled or not. A free text comment on each question was provided by the marker to augment these automatically generated comments. The automatically generated and free text evaluation was then mail-merged into a Word document that also contained some generic comments on the assignment. Each student received these three types of feedback (Excel-generated comments, marker comments and generic assessment comments) on a single sheet in time for it to be helpful to them in preparation for their traditional statistics examination.

16.3 Evaluation

Our approach to evaluation works on a range of levels and through various instruments, identified below.

16.3.1 Examination performance

In order to test whether the inquiry-based teaching and assessment improved statistical skills, examination performance was compared before and after the IBL project was introduced. The relevant examination component presented four questions to be completed in approximately one hour and twenty minutes. Typically, the questions gave a brief explanation of an experiment, some SPSS output providing descriptive and inferential statistics, and asked a number of sub-questions about analysis and interpretation. In the year before the IBL activity was introduced, the mean examination mark from 125 students was 64% (standard deviation 8.8%). In the year that the IBL task was included, the mean was 71.2% (standard deviation 9.5%) from 102 students. An independent samples t-test gave a statistically significant result ($t(225) = 5.9$ $p < 0.001$), indicating that statistical skills improved between years. It must be noted that the IBL activity was introduced within a major course overhaul in which set textbooks and lecture materials were revised and presented by a new member of staff. The examination was also set and marked by different lecturers in the two years. The differences in examination marks may therefore reflect other factors than the introduction of the IBL activity.

16.3.2 Lecturer reflection

This activity and assessment has a number of features that were novel locally to teaching and assessing statistics. Previously statistics had been assessed via traditional examination at first year. Assessment in the small group project provided a number of advantages. One advantage was that competence in using computer statistical packages was included in the assessment. Examinations offer a more effective means of assessing such competence compared to coursework, as there is no opportunity for students to use unfair strategies of collusion or plagiarism, although the collaborative nature of this examination diluted the possibility for direct assessment of individuals' ability. For this reason collaborative assessments may be best used in combination with more traditional assessments. The activity reported here contributed only 10% of the course mark, leaving room for such a combined approach. There seem to be a number of additional advantages to a collaborative assessment.

Collaborative assessment was introduced to the assignment because research is usually a collaborative process in academic psychology. This is consistent with the standard conceptualisation of IBL as a form of student-led active learning that positively models disciplinary research practices (Kahn and O'Rourke, 2005; Prosser and Trigwell, 1999). The postgraduate tutor was included in the collaboration to provide an expert resource, as methodological experts will often be available for consultation in real-world psychological research. It is speculated that this had a number of benefits for students. First, the collaboration gave a sense of shared responsibility that served to reduce anxiety. The expertise of the tutor also helped to ensure that all groups produced a reasonable solution to most questions, which may increase student confidence with statistics. As noted earlier, anxiety about statistics is a major problem in undergraduate psychology courses.

A further advantage of the tutorial programme was that, during the assessed session, there was an opportunity for the students to learn about statistical analysis. The students were able to learn by observing their colleagues' approach to the session, from the guidance provided by their tutor and from comments provided on their scripts which were returned after marking. All aspects of the tutorial task and its assessment were integrated, ensuring that student learning was both relevant and constructively aligned with the objectives of the tutor and the module as a whole. This is an approach that Biggs (1996, 2003), among others, has argued facilitates more effective student learning. Reinforcing the holistic nature of the approach, non-assessed activities of a similar sort were included earlier in the course. Students may have engaged in the non-assessed sessions to a greater extent, given that they knew a similar assessed activity would follow. It is also believed that the course mark contribution provided increased motivation for students to engage with the analysis during the assessed session itself.

16.3.3 Student evaluation

A number of questions about this activity were included in the department's usual round of feedback collection. This showed that 89% of respondents agreed or strongly agreed that the activity had improved their skills in formulating research questions. A small number of students complained that their team-mates had not contributed equitably to the task and felt it was unfair that the whole group received the same mark. In future it may be possible to ask students to rate the contribution of their team-mates and use these to weight the individual's mark within the group. This approach, although commonly used (Biggs, 2003), does not necessarily eliminate the problems of inequitable contribution. Students happy to complain informally about 'carrying' their colleagues through group work are often reluctant to mark them down, when given the opportunity (Race, 2001).

The current approach was adopted as the assessment was designed to mimic academic research collaboration as closely as possible. Inequitable contribution to group projects is likely to feature in many such research collaborations. Answers to students who raised this query highlighted the fact that inequitable contribution may be involved in professional activities they undertake in future, and so the opportunity to develop coping strategies in the relatively benign environment of the university would be to their advantage. The quantitative student evaluation indicated that 71% agreed or strongly agreed that their collaborative skills had been improved by the activity and 63% agreed or strongly agreed that their negotiation skills had improved. Students were reassured that the assignment contributed 10% of the module mark and that group marking would not be employed in assessments that contributed to their final degree classification. Despite some isolated complaints, the majority of students were positive about the collaborative aspects of the project.

16.3.4 Tutor feedback

Tutors informally reported a number of problems with the analysis session. Most importantly, they noted that ten minutes was not sufficient for each question meaning that students were put under too much time pressure. In future presentations only three questions will be included in the assessment, two specifying which variables to analyse and one asking students to generate their own question. Tutors also reported several logistical problems in room set-up and equipment availability. While it should be possible to overcome these issues with good administration, the practical burden of organising a large student cohort into groups of four to five in separate rooms with two laptop computers in each should not be underestimated.

A more substantive problem that tutors noted was that they were unsure how much help to give students with the analysis. Allowing the tutor to act as facilitator rather than examiner is desirable for a number of reasons: as noted above, it closely mimics the situation of a professional researcher, where expert statistical advice is often available; it allows some control over the students' work, to ensure they did not go too far wrong; it also provides the students with some reassurance that the task could be completed successfully and means that the session could serve as a learning opportunity as well as an assessment. For future presentation the following tutor guidelines have been prepared:

- Make sure the students run the analyses for the two explicit questions asked. If they cannot generate a solution themselves then ask students questions to try to help them decide on an answer. You should try to give less input in the write-up but make sure they do not get entirely stuck.
- Let students formulate their own question for the free question, with only very minimal help if they look like they have reached an impasse. Once they have agreed on a question, you can facilitate their selection of a test to provide an answer, but again give them little help in writing it up.

Tutors were generally positive about the experience of running the tutorials, which they appreciated gave them the opportunity to use their own research expertise in their teaching and to prepare parts of the material themselves.

16.3.5 Progression at level two

This project aimed to improve students' research skills for their final-year project and beyond. As such it formed part of an integrated programme of IBL research methods activities in the first two years. During the second year the design and analysis activity is expanded into a full piece of coursework with less input from tutors. During a laboratory class students working in groups of four choose their own topic of research, then search the literature using online bibliographic databases to learn about current developments in that area. Within the laboratory class they work as a team to develop a research design based on the existing literature. After this class, students work individually to write up their design as a research proposal. Data sets are generated for each group and each member receives a different sample from this population, then analyse and write up their results individually, as a piece of coursework. This activity is designed to build on the first-year project, allowing students greater independence to develop research skills, while still offering more structure and support than is available to them in their final-year empirical dissertation.

16.3.6 Utility of inquiry-based assessment

This project suggests that IBL has particular usefulness in a number of areas for teaching and assessing statistical learning and developing research design skills in undergraduate psychology students. The integration of a degree of student independence to the inquiry activities, the close collaboration with postgraduate tutors and fellow students in small groups and the increased sense of relevance given by allowing students to choose their own questions all positively impacted upon student engagement, even enjoyment. As was noted above, the innovations in this module form part of a broader project to embed inquiry – specifically research skills development – across all three years of the psychology curriculum. This aligns with current thinking on IBL at a curriculum design level and with the approaches taken in a number of other subject areas with which CILASS has engaged. In such projects there has been an emphasis upon supporting students through the independent and collaborative learning process and the development of baseline skills (for example, in research), from which students can then move on to more independent and advanced work at higher levels (Wood and Levy, 2009).

The inquiry task and process were established at an initial tutorial, where the students were given the opportunity to choose their topic from the list established by the postgraduate teacher. This tutorial was relatively tutor-led. This is wholly appropriate, given the level of the students, the difficult nature of the material and the strong possibility that they would establish unworkable research designs without well-structured support. The second year development of the activity is designed to give the students greater flexibility in their choice of research question and the process by which they follow them. This greater independence is appropriate at this level, when students are more familiar with the subject.

The issue of facilitation, that is, the degree of support and direction to give to students, figures highly in the literature on IBL and problem-based learning (Hutchings, 2006; Savin-Baden, 2003). As with the degree of open-endedness of the inquiry task and process, the extent to which individual tutors direct, support and monitor those processes is dependent upon student level, intended learning outcomes and disciplinary approaches. The reported reactions of tutors on this module to this learning approach is entirely in line with the literature; tutor anxiety over the issue of facilitation is also commonplace and it is important to consider this when offering support and advice to those teaching in this manner, especially those who are inexperienced in inquiry approaches (Kahn and O'Rourke, 2005; Goldring and Wood, 2007).

The activity also served to strengthen links between teaching and research. As the tutorials were structured around topics of interest to the postgraduate researcher, the students were introduced to topics that are being actively researched in the department. By encouraging the tutors to engage explicitly with IBL pedagogy in the context of their personal research interests, the approach taken on this module would seem to offer an opportunity for strengthening research-teaching linkages. Such links may have important benefits both for student learning and for staff teaching and research (Brew, 2006; Jenkins and Healey, 2005).

Part D

INDIVIDUALISED ASSESSMENT

17

Individualised assessment in statistics

Neville Hunt

17.1 Introduction

Individualised assessment is where each student in a class group is allocated a different task, supposedly of equal difficulty. The principal motivation for individualised assessment is to deter academic dishonesty, namely plagiarism and collusion. A secondary consideration is to make the assessment more rewarding for the student and thereby increase engagement with the learning process.

Individualised assessment is not new. Twenty years ago the author recalls assisting on a statistics course where one of the assignments required each student to collect their own small set of bivariate data and carry out a regression analysis. These students demonstrated commendable imagination and the fruits of their labour provided the lecturer with a rich source of examination questions for many years. (I do not recall the lecturer ever being accused of plagiarism!) More than forty years ago it was commonplace for students to conduct simple experiments in the 'statistical laboratory' and for each to write up their results as a 'lab report' (Murdoch and Barnes, 1974). Given the random nature of most of the experiments it was inconceivable that two students could obtain the same results, so it was futile to collude. Increased student numbers make this type of assessment less common nowadays, because of the resource implications. What *is* new is that technology has made it easier for the teacher to set and mark individualised assessments. In a survey of staff at 23 universities in the United Kingdom, the PiSA project (Bidgood *et al.*, 2007) discovered a thriving cottage

Assessment Methods in Statistical Education Edited by Penelope Bidgood, Neville Hunt and Flavia Jolliffe

industry of homespun computer-based individualised assessment systems. Some of these will be discussed here.

The ultimate individualised assessment is the extended individual project. This is where each student is allocated a completely different problem to work on. One student may be analysing baseball results whilst another is forecasting stock market prices. Such an approach is very resource-intensive, requiring individual supervision and substantial marking time. In a United Kingdom degree course a student would normally only expect to meet such an assessment once, typically in their final year. Although this can never be the normal assessment instrument, it remains an ideal to aspire to and a benchmark by which to judge other individualisation schemes.

17.2 Plagiarism

Plagiarism in this context is the act of presenting someone else's work as one's own, whereas collusion is working together with intent to deceive the assessor. Although no official figures are available, hearsay evidence suggests that there has been a marked increase in reported incidents of plagiarism and collusion in recent years. This does not necessarily mean that plagiarism itself has proliferated; it could simply be that teachers are now more aware of it and institutions have formal procedures for dealing with it. However, explanations for a proliferation are not hard to find.

In the United Kingdom the increase in plagiarism has gone hand-in-hand with an increase in class sizes. There was a time when classes were smaller, and two students in a class of 10 had no hope of their collusion going undetected. Now, on the other hand, two students in a class of 300, where assignments are marked by many different teaching assistants, may consider the odds quite favourable. The personal element is also a factor. In a class of 10 the lecturer knows each student personally and to try to deceive the lecturer seems like betrayal, whereas in the class of 300 a student may feel completely anonymous and may never have had a conversation with the lecturer. One might draw a parallel with the shoplifter who would not dream of stealing from an acquaintance yet has no such qualms about a major high street store.

Plagiarism and collusion have been assisted by advances in technology, particularly in electronic communication. It is now so much easier to plagiarise, drawing on trillions of Internet sources. It is also so much easier to collude. Electronic documents can easily be exchanged and subtle alterations made using sophisticated computer editing tools. Through online social networking sites one may even encounter a former student who was set the very same assignment several years previously and just happens to have kept the model solution. Additionally, students have access to a range of online consultants offering perfectly legal ghostwriting services.

Many lecturers have realised that time invested in designing an assessment strategy that avoids plagiarism, may reap the reward of time saved on detecting, reporting and prosecuting offenders. The reaction of some has been to abandon the traditional take-away statistics assignment in favour of supervised examinations, possibly computer-based, or open-book, or based on a seen scenario. Yet others have turned to the online plagiarism detection service Turnitin (www.turnitin.com) to provide the required deterrence. When used over a period of several years, Turnitin has the added advantage of being able to detect collusion not just within a cohort but also between cohorts past and present. The limitation of Turnitin is that it is really only usable for written reports submitted electronically.

17.3 Individualised data collection

As indicated in the introduction, a popular method of individualising an assessment is to require students to collect their own data, either primary or secondary. This has the advantage that students can choose a data set that appeals to them, and they therefore become stakeholders in the assignment with a personal interest in the conclusions of any analysis. This approach is not without its problems.

First, there is the danger that a student may collect completely inappropriate data, rendering any subsequent analysis worthless. To avoid this pitfall the lecturer must either provide a very detailed specification of the data to be collected, or insist on approving students' data sets before they begin any analysis, as illustrated by Mashhoudy in Chapter 21.

Second, there is a risk of plagiarism. A student may locate a worked example from the Internet or from an obscure textbook, allowing them to plagiarise not only the data but also an exemplary analysis. Insisting that students cite their data source helps to prevent this. If the lecturer has set the same assignment before, a student may re-submit the work of someone from an earlier cohort and, unless the lecturer keeps copious records, this might be very difficult to prove. One way of combating this is for the lecturer to require data to be current or very recent, not more than a year old. Of course, a student may completely fabricate primary data but, as lecturers know all too well, it is quite difficult to construct fictional data that have the required properties and are not 'too good to be true'. Ultimately it is the fraudster's loss, since the analysis will be a chore with no meaning and no pleasure of discovery.

Third, the marking burden on the lecturer is undoubtedly increased. Every student's work is unique. Unless the class is very small, the lecturer will almost certainly not be able to check the numerical accuracy of each student's analysis. Blatant errors can still be identified. The author remembers a student submitting a regression analysis of the recorded mileage of cars of different ages, in which he had mistakenly predicted the mileage of a car 100,000 years old – not understanding the scientific notation used in the computer output, he even commented that he thought his prediction was very accurate! Removing the focus from the accuracy of calculations may be beneficial, forcing students to appreciate that it is how they interpret the numbers that really matters. It is not uncommon for a lecturer to award a mark of 9/10 to a student who has conducted a hypothesis test immaculately – apart from drawing the wrong conclusion! This sends entirely the wrong message about the relative importance of calculation and interpretation.

17.4 Student-driven individualisation

Another popular method of individualising assignments is to require students to each select a unique subset of a large data set. Jolliffe (2003) suggests that the choice of sample can be left to the student but Hunt (2007a) favours utilising the student's ID number, which is usually at least four digits long. Some care is needed in case particular digits represent the year of entry, but otherwise each digit of the ID number can be treated as effectively random, allowing the statistics lecturer to devise all kinds of clever schemes to allocate a different sample to each student. For example, a student with ID number 4921 could be required to delete rows 49 and 21 from the master data set – but the voice of bitter experience warns us not to forget the possibility of 00!

Taking a random sample of the rows of the data set is not in itself much of a deterrent against collusion if the same task is set. For most students the conclusions will be broadly the same, with only minor variations in the numbers. If the task can be completed using a spreadsheet, once one student has performed the analysis, a friend can simply paste their data over his and the whole analysis automatically updates. If the analysis is conducted by a command-driven computer package, again, once one student has written the program any of his friends can easily re-use it. A more sophisticated approach is to use the student's ID number to select not only random rows but also random columns of the data set: in a multiple regression analysis there might be a common y-variable but each student is allocated a different selection of potential x-variables. Some lecturers have gone further and used the ID digits to randomise the tasks themselves. For example:

> Replacing the letters WXYZ by the four digits of your student ID number, answer the following:
>
> Predict the mean selling price of a car that is (1+W) years old and has X previous owners.
>
> Calculate and interpret a (90+Y)% confidence interval for this mean.
>
> A car that is Z years old is being offered for sale at £300(W+X+Y). Discuss whether or not this is a bargain.

17.5 Computer-driven individualisation

Modern computer technology has allowed some lecturers to automate the various approaches outlined above. Computer software is now being used for some or all of the following:

1. to individualise the assessment for each student,
2. to deliver the assessment to each student,

3. to receive completed work from each student,
4. to mark each student's work automatically,
5. to release the mark to each student, confidentially.

A variety of different software environments are being used. Simonite *et al.* (1998) and Hunt (2005) describe how the standard mail merge facility within Microsoft Office can be used to create individualised Statistics assignments. The author (Hunt, (2007a, 2007b)) uses Visual Basic for Applications (VBA) macros within Excel to create his ISCUS (Individualised Statistics Coursework Using Spreadsheets) system. Spencer, in Chapter 18, also uses VBA macros in Excel but within the existing familiar interface of his institution's virtual learning environment (VLE). In Chapter 20, Simonite and Targett use purpose-built Visual Basic programs to create a system that manages all five aspects above. The DRUID system (Davies and Payne, 2001) uses web-based technology. The key to success for all these systems is simplicity. Many students nowadays are distance learners carrying out much of their work at home. Any system that requires them to have the latest version of some proprietary software will generate complaints. Many statistics lecturers are not computer boffins. Any system that requires great technical expertise from them will simply not be used. Longevity is also an issue. We can all recall projects that developed brilliant products that were obsolete within two years of completion. To be fair to Microsoft, it is worth noting that many Excel 4 macros written by the author in 1995 are still functioning unmodified in the current 2007 version of Excel.

For small classes, it is feasible to produce personalised paper copies of assessments; for larger classes the printing and distribution becomes quite onerous. The current mail merge facility in Microsoft Office allows email merge, where the individual documents are distributed by email rather than in paper form. Spencer sends each student a personalised spreadsheet by email attachment (see Chapter 18). Simonite and Targett store the individual assessments as HTML files on a web server and send students an email containing a hyperlink to their personal file (see Chapter 20). email distribution is undoubtedly more convenient, although some provision needs to be made for students whose mailbox was full at the time of the mailing. The author's ISCUS system overcomes the distribution problem by creating a single assignment generator spreadsheet, which is stored on a central network that all students can access. Each student then opens the spreadsheet and enters his/her ID number before running a macro that creates for them a unique spreadsheet containing their task and/or data. Although the data are supplied on a spreadsheet, they do not need to be analysed in Excel since most statistical software packages accept spreadsheets as data files.

The principal benefit of a computer-driven individualisation system is its potential for automated marking. The author's ISCUS system makes no provision for electronic submission by students and hence cannot offer automatic marking. The lecturer can create an answer-generator spreadsheet that produces numerical outputs for each student's data, but this is merely an aid to manual marking. In Simonite's system students submit their answers via a web page and the lecturer runs a marking program that automatically marks the work and creates an individual feedback file for each student, which they again access via a web page. Similarly in Spencer's system students submit electronically via the VLE. A further Excel spreadsheet then harvests their answers, marks them and provides relevant feedback. A particularly impressive feature of this system is that it allows marks to be awarded for correct work following on from an earlier error.

In Chapter 19, Stirling describes his CAST system where students access and complete randomised Statistics exercises online. Most universities now operate a web-based VLE that can be used to conduct computer-marked online assessments, including secure recording of marks. The initial investment of time spent creating the question bank is repaid by the savings in test administration and marking. As the PiSA project (Bidgood *et al.*, 2007) discovered in its survey, such tests are not necessarily supervised:

> Several lecturers reported using the assessment tool within their institution's virtual learning environment (e.g. Blackboard) to create randomised computer-marked online quizzes which students must complete on a regular basis. Although these were typically unsupervised, it was felt that each quiz carried such little weight in the overall assessment that the advantages gained by student engagement outweighed any possible risk of collusion. Although this type of assessment fails to address some important learning outcomes for Statistics students, it was seen as a pragmatic response to very large class sizes.

The randomisation within tests can take several forms. In multiple-choice tests some VLEs randomise the order of both the questions and the different responses as a way of deterring cheating, as discussed by McLeod *et al.* (2003). Others allow the creation of question sets containing multiple copies of a question, each with different values for some numerical parameters. Each student who sits the quiz receives a different, randomly selected question from each set. This facility is particularly useful for formative assessment, where students can be encouraged to keep retaking the quiz until they have mastered it. The exercises set by CAST have so many random features that they can quite reasonably be used for summative assessment even after students have used them for formative assessment.

17.6 Discussion

As Hunt (2007a) points out, individualisation of assessment can create problems of fairness. Whilst all students may have apparently been set the same task, the particular data set allocated to one student may make the task more difficult than for other students. For example, there may be an outlier in the data set that only certain students have the misfortune to be allocated. Missing data can cause similar problems. With experience the lecturer will learn to look out for these things and either eliminate them or ensure somehow that every student is allocated the difficult data. The author once set a regression assignment where, although there was a significant correlation between the two variables in the large source data set, many of the small samples allocated to students showed no correlation at all, rendering parts of the assignment trivial for them. Any consideration of sampling distributions should have warned the author that for a characteristic to be evident in every sample it must be exceedingly strong in the population.

In individualised assessment there will always be a tension between the desire to set tasks that can be easily marked and the desire to set tasks that are educationally beneficial. Undoubtedly the move towards individualised assessment has encouraged the setting of closed tasks with numerical 'answers'. Some see group work as the solution, allowing more realistic and challenging tasks to be set without increasing the marking burden – 25 group posters as opposed to 100 individual reports. In group work, collusion can be managed. Although there are difficulties in assessing the contribution of an individual to a group effort, these are not insurmountable, as Hannigan shows in Chapter 15. Moreover, group working provides a more realistic preparation for employment. In the real world whilst it is conceivable that a manager might issue the same task to several employees and ask them to produce an analysis independently, perhaps to elicit a variety of ideas, it is far more likely that the task would be allocated to a team to work on together. It would indeed be ironic if the ultimate cure for collusion was not individualisation but collaboration.

18

An adaptive, automated, individualised assessment system for introductory statistics

Neil Spencer

18.1 Introduction and assessment aims

This chapter concerns the use of individualised computer-assisted assessment carried out as part of a module for first year undergraduate students at the University of Hertfordshire's Business School studying for a degree in marketing. The module's learning outcomes include:

- being able to apply appropriate quantitative techniques to manipulate and solve a range of business problems, and
- being able to use computer software to solve quantitative problems.

When considering an appropriate assessment strategy for this module, it was decided to adopt an approach that was similar to the way in which marketing students would undertake quantitative work in future employment. It was thus decided that it would not be appropriate to have an examination as part of the process but rather to give students a series of tutorial sheets. They would

Assessment Methods in Statistical Education Edited by Penelope Bidgood, Neville Hunt and Flavia Jolliffe

be given relatively tight deadlines within which to complete and submit the work, and thus have to plan their time accordingly. These tutorial sheets would contribute 70% of the marks available for the module, with a piece of group work contributing the other 30%. Another advantage of asking the students to undertake a number of tutorial sheets would be that it would encourage them to engage fully with the learning required for the module at an early stage of the semester. This was considered to be especially useful for the development of their learning at university, since the module took place in the first semester of their studies.

Because the students would be undertaking the work for the tutorial sheets unsupervised, the issue of potential plagiarism had to be addressed. It was decided that each student should receive a slightly different tutorial sheet. So that the assessment experience was the same for each student, they would be asked to perform similar tasks in any particular tutorial sheet but the data upon which the calculations were based would vary from student to student. Although students would be able to discuss with each other how the questions in the tutorial sheet might be answered, each would have to undertake the calculations specific to their own tutorial sheet. Any discussions that they had with each other would thus be regarded as a useful part of the learning process.

In order to create individualised tutorial sheets for a significant number of students (ranging from approximately 90 to 140 over recent years), it was recognised that a computer-based system would be needed. It was also recognised that the task of marking the answers supplied by the students would be considerable and time-consuming. If feedback was to be given to students in a timely fashion, a computer-based system of marking would need to be utilised.

18.2 Other work on individualised assessment in statistics

Over recent years, work has been undertaken at a number of Higher Educational establishments to produce individualised assessments for students studying statistics. As Bidgood *et al.* (2007) report, these developments have been taking place largely independently of each other and only a limited amount of the work has found its way to publication. Much of this has focussed on the use of computers to create different data sets for each student on a course or asking students to answer questions that contain different parameters (Jolliffe, 2003). The system described by Hunt (2007a) can produce individualised solutions (to accompany the individualised assessments) for tutors to use when marking assessments manually. Others working in this field (e.g. Simonite and Targett, see Chapter 20) are moving towards systems that allow assessments to be marked automatically once the students have submitted their answers electronically, and that is the focus of this chapter.

18.3 Russell's system

Having decided on the style of the assessment that the author wished to implement for the module discussed earlier, the author was made aware of work undertaken elsewhere in the University of Hertfordshire by Mark Russell. He had developed a system similar to that required for the author's module and was keen to cooperate with others who wanted to use the system. His system is discussed in detail in Russell (2005), but the essential elements are as follows.

For each number required in a tutorial sheet, a random number is created by Excel. Various bounds are placed on the random number so that it will make sense in the context of the question in which it is to appear. This may include minimum and maximum values and whether the number is an integer or has a specific number of decimal places. These random numbers are stored for each student.

A template for the tutorial sheet is created in Word, with mail merge 'fields' being used wherever a number that will change from student to student occurs. The stored random numbers in Excel are then used to create the individual tutorial sheets with the mail merge facilities of Word.

The individual tutorial sheets are loaded onto the University of Hertfordshire's virtual learning environment for the students to download. They undertake the work required in their own time.

Once the students are ready to submit their answers, they go to a computer which has a link to a 'Data Gatherer' program. They have been provided with a username and password, and use these to gain access to a series of boxes into which they must type their answers to the questions on the tutorial sheet. They then click on a button to submit their answers and are given a 'receipt' number as proof that they have submitted answers.

Once the deadline for submission of answers has passed, the tutor downloads the answers from the Data Gatherer and transfers them to Excel.

The marking of the answers is undertaken in Excel. The correct answers are known and can be checked against the students' answers. If a student's answer almost matches the correct answer but not exactly, then partial marks can be given. The tolerance allowed for this near-match can be determined on a question-by-question basis. If the answer to one question is used in the calculation of a subsequent question, then the student's answer to the first question can be used, even if it is wrong. In this way a student who uses a correct method using an incorrect answer from a previous question will still get the credit deserved.

Using Visual Basic for Applications (VBA) programming in Excel, the marks are relayed to the students by email.

This system was used by the author for the first year that his module took place. Feedback from students was positive. In comparison to a previous incarnation of this module (which had an examination as a major part of the assessment process), tutors experienced a greater engagement with the module by students and the final marks achieved improved.

18.4 Issues raised by using Russell's system

Although the use of Russell's system was successful, the potential for its refinement was noted by the author. Areas where the system did not work as well as it might have been expected were as follows:

(i) A significant number of students struggled with the Data Gatherer system for submitting their answers. Having a separate username and password specific to this system confused some students and led to them submitting their answers late or not at all. Russell had not encountered the same level of problems with his students, but as they were enrolled for degrees involving engineering, it might be hypothesised that they were more capable of coping with technological challenges than students studying marketing.

(ii) Students had to go to one of the University's Learning Resources Centres in order to use a computer that had a link to the Data Gatherer program. This was a general inconvenience to students, but more importantly meant that students who returned home for vacations were unable to submit any work while away from the university. At that point in time, at the end of the first time the module had taken place, Russell was planning to develop a web-based Data Gatherer. However, it was clear that it would not be ready in time for the next time the author's module was to run.

(iii) On many occasions students lost marks, not because they had done the work incorrectly, but because they had made mistakes in copying their answers to the relevant boxes of the Data Gatherer program.

(iv) A student who had undertaken the work correctly in all but one aspect was awarded the same zero marks as a student who had not attempted the question. In a traditional examination, a marker would be able to see that a student had made a relatively small error (e.g. using a variance instead of a standard deviation in a calculation) and give the student some marks for getting the rest of the method correct. The marking system used was not capable of discovering this and awarding partial marks, because the difference in the answer obtained by the student might be considerable in size. The tolerance allowed for an answer to be regarded as a near match could not be wide enough to encompass this possibility.

Enthused by the success of the first implementation of Russell's system, and understanding that Russell did not have the resources to amend his system to suit the needs of the author's module, the author decided to further develop the system himself. At this point the development of Russell's system and that of the author diverged.

18.5 Further developing the system

To overcome the difficulties that the Data Gatherer caused some students, the author decided to make use of another system for submitting work that is available to students at the University of Hertfordshire. In common with many Higher Education institutions in the United Kingdom, the university has a virtual learning environment (VLE). Its version, Studynet, allows students to submit electronic work online, providing tutors have set up appropriate sections of the module's Studynet website to accept it. Although students have to enter a username and password to access Studynet, they use this system for so many other University services (e.g. email, course materials, news, receiving assessment results), as well as submitting assignments, that they quickly get used to it.

In order to make use of this facility in Studynet, students would have to submit a file containing their answers instead of putting them in boxes in the Data Gatherer. There would also have to be some facility for extracting the student's answers from whatever files they submitted on Studynet. As the students were being encouraged to use Excel as a tool with which to carry out the calculations needed for the tutorial sheets (including the use of Excel's built-in functions, where appropriate), it seemed natural to use Excel as a way of students submitting their answers. To facilitate this, it was decided that rather than supply students with individualised tutorial sheets that had been created by mail merge in Word, they would be emailed an individualised Excel workbook. The workbook would contain the questions that they had to answer, but also have specific cells in which the students should type their answers. Students would then submit the completed Excel workbook via Studynet.

Once the deadline for the assessment had passed, the workbooks could be retrieved from Studynet onto the author's computer, ready for marking to take place. To extract the answers from each workbook, it was necessary to know exactly where in the Excel workbook the students had entered their answers. Figure 18.1 shows an example of how part of a student worksheet appears – the

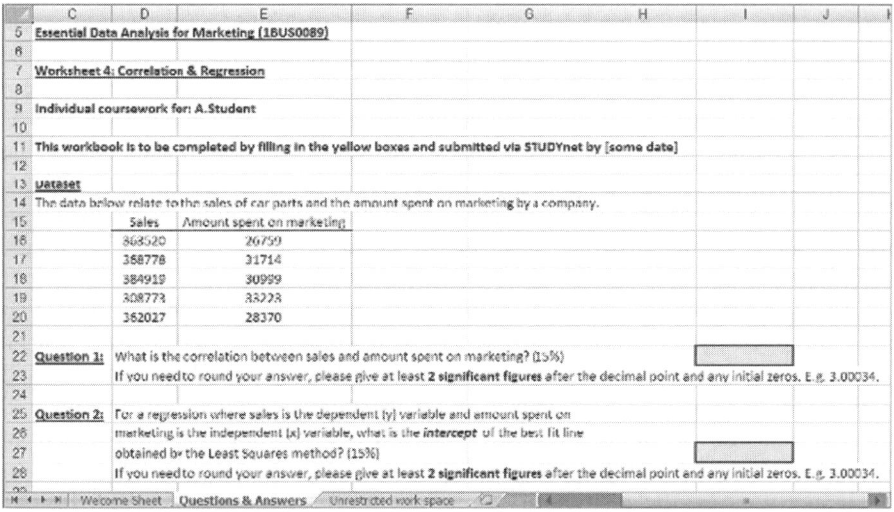

Figure 18.1 Example of part of a student worksheet.

shaded cells are where the students are required to put their answers. Although it was made clear in the appropriate worksheet that students should put their answers in certain cells, it was also thought wise to password-protect the worksheet so that students could only edit the cells where they were to place their answers. The facility to copy and paste the worksheet questions to another workbook was also disabled, to avoid students creating and then submitting workbooks that were not formatted in the manner expected.

A useful facility that existed with Russell's Data Gatherer system was that if a student tried to submit a set of answers that were incomplete (i.e. had empty boxes where answers were meant to be placed), the student would be warned of this and given the opportunity to return and complete the boxes before submitting the answers. To duplicate this facility in Excel, macros were written in VBA and distributed along with each student's individualised Excel workbook. When a student tries to save the workbook or exit without saving, the macros check whether any of the cells meant to contain the answers are empty and warns the student, giving them the opportunity to return and place answers in the appropriate cells.

The above amendments to Russell's system were useful in avoiding the difficulties students encountered listed in (i) and (ii), above. The author was aware that in order to overcome difficulty (iii), students may be tempted to copy and paste a formula from another workbook into the workbook that they were to submit for marking, or simply calculate the answer with a formula within the cell required to contain the answer to the question. This could cause problems if the formula referred to another workbook that was on the student's computer but which would not be present on the tutor's computer. From the student's point of view, the correct answer would be displayed and they would happily submit the Excel workbook. However, no marks could be awarded because Excel would not be able to find the linked workbook when it came to being automatically marked on the tutor's computer. It might be possible for the tutor to look at the formula and see that some parts were correct, but it would be impossible to see precisely what calculations the student had made and it would also be impossible for the marking to be undertaken automatically by computer. Another problem with formulae was that it opened up the possibility of students plagiarising work. If it was possible to construct a formula to answer a question in such a way that it could refer to particular cells in a worksheet rather than particular numbers, then it would be possible for a student to distribute the formula to their acquaintances and each student would have their own individualised answer calculated automatically and correctly by the formula.

To overcome these problems initially, macros were written to check that answers given by the student did not contain formulae. If any answers with a formula were found, the student would be forced to change them before being allowed to save the workbook. However, this meant that students still encountered the issue mentioned in (iii), above. They still had to copy their answers from one place to the worksheet cell where they were required to place the answer. To help with this, the restriction on the use of formulae was relaxed. Instead of rejecting all formulae, the macros were altered so that only formulae that referred to other workbooks were rejected. To overcome the potential problem of students distributing formulae that would give correct answers for everyone, each student's workbook had a random number of rows and columns added to the top and left of the worksheet containing the tutorial sheet questions. These rows and columns were hidden in Excel so that the appearance of the tutorial sheet was the same for every student but the actual cells where they needed to place their answers was different from student to student. This means that if one student were to copy a formula from another, they would have to change the cell references in the formula to match the arrangement in their own workbook. By doing this they would effectively be undertaking a learning experience. If they were successful in altering the formula, they would be demonstrating an understanding of how to answer the question and thus be deserving of the marks that would be awarded.

An additional benefit of allowing students to use formulae is that in so doing they are demonstrating an ability to use Excel to undertake the required calculations. As the students are encouraged to use Excel in this way during the module, this can be considered to be an additional demonstration of their achieving the learning outcomes for the module. At the time of writing, the author is considering further changing the macros in the future in such a way that students are forced to use formulae to answer the questions. Care would have to be taken in this venture to avoid giving credit to a student who has used a formula such as '=34 + 1' when the intention is to give credit for using one specific function from those available in Excel.

The use of macros in Excel does introduce a difficulty. For reasons of system security, most installations of Excel do not automatically allow macros to be run when a workbook is opened. Macros may be disabled entirely or disabled until the user specifically allows them to be run. If students were able to have access to their tutorial sheets without allowing macros to run, many would undertake the work without them running and would not be able to benefit from the protection against blank entries and formulae linked to other workbooks that the macros provide. The implications of students submitting answers that are blank or that are linked to workbooks that the tutor does not have on his or her computer would cause wasteful administrative problems that would be avoided if the macros were allowed to run.

To overcome this, when the students open up the workbook containing the tutorial sheet, they are not faced immediately with the tutorial sheet but with a 'welcome' page. This contains a button that they must click and, once they do so, a macro is run that reveals the tutorial sheet that has been hidden from view until that point. Thus, in order to view the tutorial sheet, they must run a macro, and in order for them to run a macro, they must have allowed all macros connected with the workbook to run. As well as the button they need to click, the welcome page also contains a detailed explanation of what they must do in order to allow the macros to run. As well as these explanations, the students are also sent a non-assessed tutorial sheet in the first week of the module. In the module's first workshop session, they are instructed to practise opening, completing and submitting this non-assessed tutorial sheet. The small number of students who miss this first workshop are offered instruction in the process in subsequent weeks.

18.6 Marking students' work

Once all the workbooks submitted by students have been downloaded and the answers extracted, the marking must take place. Russell's system already had a marking procedure that not only checked if the correct answer was given, but also allowed marks to be given for answers which were within a certain distance of the correct answer (i.e. the student's answer is 'near enough' to be deserving of some marks). Russell had found this procedure to be suitable for the engineering discipline in which he was operating, but in statistics a small error in the answer may be indicative of a conceptual error on the part of the student. For example, consider the question:

> A survey asked for the ages of 22 customers in a shop. The sample mean was 34.04 and the standard error of the mean was 0.726. Which value from the t-distribution should be used when constructing a 95% confidence interval?

(Other questions associated with this scenario might also be asked). The correct answer is obtained by taking the 97.5% point of the t-distribution with 21 degrees of freedom, namely 2.414. Full marks could also be given for the 2.5% point (−2.414) if a student gave this answer. A student who used 22 degrees of freedom instead of 21, but still correctly used the 97.5% point (or 2.5% point), would obtain the answer 2.405 (or −2.405). One would not wish to give the answer obtained using 22 degrees of freedom full marks, despite the fact that it is identical to the correct answer to two decimal places. Instead one would wish to identify the wrong answer as being wrong, but also as being worthy of some marks, as it is the correct calculation when n is used instead of $n - 1$ as the degrees of freedom. For this question, another method that is wrong that may be considered worthy of partial marks would be using the 5% or 95% points of the t-distribution.

To cope with this and also address the issue mentioned in (iv), above, required the system to be amended in such a way that not only is a student's answer checked to see if it matches the correct answer or if it is 'near enough' to this answer, but that it also checks a range of alternative answers. It is impossible to be able to anticipate all the ways in which a student might get an answer wrong, but it is possible to guess a good number of such ways. The amended marking scheme can then check to see if the student's answer matches any of these alternative wrong answers. If a match (or 'near enough' match) is found, then marks can be awarded as appropriate. When an email is subsequently sent to the students giving them the results of marking their tutorial sheet, additional feedback can be given. For the example above, the feedback might be: 'I reckon you have incorrectly used n degrees of freedom instead of n − 1 when obtaining the value from the t-distribution. You have still been awarded some marks for getting the rest correct'.

The student thus receives formative as well as summative feedback. The marking system was also programmed to look for, and award marks for, answers that are wrong by a factor of 10 because students have mistakenly entered too many or too few zeros between the decimal place and the first significant figure in their answer.

Because of the difficulty of guessing what students might do wrong, the emailed feedback also encourages them to challenge the marking system. If students can demonstrate that they got their wrong answer by doing something that was deserving of at least some marks, then the marking system can be programmed to look for these new 'wrong methods' and award marks appropriately. Although encouraging students to challenge marks potentially might have resulted in a flood of students trying to gain extra marks, this has not been the author's experience. Students have appeared to appreciate the openness with which the marking has been carried out. Where challenges have been made, they have been with an understanding that the marking system cannot be expected to perform miracles and guess what they might do wrong. Where mistakes in programming have, on very rare occasions, denied students marks to which they were entitled, they have been glad of a system which encourages them to point this out, reacting with relief that a solution is found rather than one of annoyance that a mistake has been made.

The biggest difficulty that remains with the marking of tutorial sheets is coping with the rounding-off of answers that students undertake. The author feels that it is not reasonable to be too strict with students who are studying for a degree in marketing when it comes to asking for a specific number of decimal places in the answers. Initially the author allowed students to round their answers to however many decimal places they thought was reasonable, but this caused difficulties when some are rounding to one and others to 12 or more decimal places. Those who are, perhaps, over-rounding are in danger of being awarded marks that they do not deserve. For example, if their calculations yield a probability of 0.144 for a question, when the correct answer should be something quite different such as 0.065, since both round to 0.1, the student may be inappropriately awarded marks for getting the answer correct. Also those who are, perhaps, under-rounding may be unfairly penalised. For example, if they incorrectly round 0.254465283 to 0.25446, they have indeed made a mistake, but one can understand that they might feel aggrieved if they are penalised when someone who has given 0.25 as an answer receives full marks.

To avoid this problem, the author moved on to asking for a particular number of decimal places (as appropriate for the question involved). This resolved the problem to some extent, but also caused an additional one. When a student does something that is wrong, but still deserving of some marks, it is not unusual for the answer that they give to be rather extreme. For example, if when conducting a t-test they use the variance instead of the standard deviation, the resulting p-value may be rather large and near to one. A student with a wrong answer, such as a probability of 0.999873, may correctly round it to three decimal places to get 1.000. It then becomes impossible to tell whether the answer given by the student is the result of rounding an answer of 0.999873, which is worthy of some marks, or the result of doing something that is deserving of no marks at all, such as somehow obtaining a probability of 1.000473 which is then rounded to 1.000. At the time of writing, the author has overcome this issue by asking for a rather non-standard style of rounding: students are required to give the answer to a specific number of significant figures after the decimal point and after any zeros or nines that immediately follow it. Although this is a rather peculiar request, the students are given illustrate examples of this alongside each question, and appear to be managing to understand and comply.

If the system is further developed to force students to use formulae in their answers (as mentioned above), then this rounding issue will largely disappear. Excel will simply store the answers to the default precision that it uses and students will not have to round their answers. For questions where the answer to a previous question is used in the calculation, students could use the cell reference for the first answer in the formula for the second question. This could be considered a backward step, as students would not learn how to round numbers correctly. However, if this was an important learning outcome for the module, it could be taught explicitly and assessed via this individualised system using programming for certain questions that disallowed the use of formulae.

18.7 Summary

We have described an adaptive, automated and individualised system for assessing student undertaking an introductory statistics module. The system is adaptive in that it can adapt to different mistakes that a student might make and still award partial marks. It is automated in that the creation of the worksheets, the marking of the worksheets and the sending of feedback is all carried out by computer programming. It is individualised in that each student will get a different set of data from which they need to compute their answers.

In summary, the 'unique selling points' of this system compared with some other similar systems are as follows:

(i) The Excel workbooks containing the work to be completed can be delivered direct to the students via email or other means.

(ii) The system can, if desired, force students to use formulae in Excel, thus potentially ensuring the attainment of a learning outcome.

(iii) The completed student work can be submitted via a VLE in exactly the same way as an essay or other piece of coursework.

(iv) The marking system can allow for students producing wrong answers using a partially correct method and still award partial marks.

(v) In connection with (iv), formative as well as summative feedback can be given.

(vi) By encouraging students to challenge if they believe they have done something worthy of marks, they will take more notice of feedback given.

The author has had great success implementing this system at the University of Hertfordshire and is willing to share what he has developed with others.

19

Random computer-based exercises for teaching statistical skills and concepts

Doug Stirling

19.1 Types of assessment

Educators often distinguish between summative assessment, where the main aims are to rank students and occasionally compare them to existing standards or norms, and formative assessment, that is designed to help students learn the material in a course (Scriven, 1967). In reality any piece of assessment lies on a continuous scale between these two extremes. The same mid-semester test may be given with no feedback other than its contribution to the student's final course mark, or it may be returned with extensive comments but no contribution to the final mark. An assessment activity's position on the summative-formative scale depends more on what happens afterwards than its actual content.

The practice of statistics requires mastery of an extensive toolbox of concepts and methods and short, fine-grained questions about specific topics are often used for purely summative assessment (e.g. in multiple-choice examinations), for purely formative assessment in exercises, or as a combination in assignments (Garfield, 1994). This chapter discusses such short questions with particular emphasis on the feedback that can be provided by computer-based exercises.

Assessment Methods in Statistical Education Edited by Penelope Bidgood, Neville Hunt and Flavia Jolliffe

The main focus of a disappointingly large proportion of students is their final mark; hence many courses include a summative component in most assessment activities. To encourage student participation in purely formative assessment activities, they must be clearly seen as essential preparation for later summative assessment. For example, the use of exercises by students in their study is increased if similar questions are known to appear in a later multiple-choice exam.

It can be argued that the main goal for students in an introductory statistics course should be the ability to collect, analyse and interpret data in different application areas and that project work involving real data should therefore be the focus of assessment (Kanji, 1979). However, whether or not short questions are a component of summative assessment, exercises do have an important role in mastering individual concepts and skills before embarking on relatively open-ended projects.

19.2 Exercises in textbooks

Exercises are a form of active learning (Meyers and Jones, 1993), in which students do more than passively listen to a lecture or read a book. They require students to think and recall information at each step and are therefore more intense than many other forms of study. Interaction through feedback about mistakes makes the learning process even more active. A well-designed set of exercises can cover a wide range of statistical concepts and skills.

Textbooks (both print- and web-based) usually include short exercises about the topics in each section or chapter, often to the exclusion of more open-ended or wide-ranging questions. Traditional exercises in textbooks give little feedback about wrong answers; at best, a brief answer is shown at the back of the book, but often this is only for selected questions and sometimes only a single number is provided as the solution. Students are often left without any indication of where they went wrong if their answer is incorrect.

Many paper-based textbooks have associated CD- or Internet-based resources that contain similar short-answer exercises and/or tests. For example, many textbooks published by the Thomson group of publishers are linked to *CengageNow* resources (Cengage Learning, 2008) that provide multiple-choice tests for their chapters. Although more information can be provided about wrong answers than in textbooks, the multiple-choice format usually limits the types of question and scope for feedback.

Several introductory statistics textbooks have been written and published for direct use in a web browser in recent years, often without payment. Although dynamic and interactive features can allow electronic textbooks to effectively teach statistical concepts, there are often no exercises, as in *StatSoft Electronic Statistics Textbook* (StatSoft, 2008), *StatWeb* (Hulme *et al.*, 2008) and *Visual Statistics Studio* (Cruise Scientific, 2008), or the questions have the same format and limitations as those in paper-based textbooks, as in *Hyperstat* (Lane, 2008), *StatPrimer* (Gerstman, 2008) and *Statistics at Square One* (Swinscow, 2008).

19.3 Computer-based exercises

Computer-based exercises have the potential to be much more flexible and effective. For example, the exercises by Perdisco (2008) include multiple-choice questions, questions requiring numerical answers and a few questions in which something must be dragged in a diagram. In general, exercises might require:

- a multiple-choice response;
- a numerical value (e.g. a guess at a correlation coefficient from a data display);
- interaction with a diagram (e.g. to 'sketch' a histogram or draw a least squares line);
- several responses of these types, perhaps in sequence.

Exercises might also provide interactive diagrams and formula templates (with text-edit boxes into which values can be typed to perform calculations) to help answer the questions.

Good feedback enhances the value of any exercise as formative assessment. In a computer-based exercise it is possible to analyse the student's attempt and give specific feedback about mistakes, perhaps tailored to common types of error. Ideally the feedback should be of the same quality as that provided by a human tutor.

19.4 Randomisation of questions

Many collections of exercises contain too few questions of any type, preventing weaker students from repeating similar questions until a skill has been mastered. A computer-based exercise can be randomly generated over such a wide range of question variations that repetition would only make answering the question easier through mastery of the targeted skill. Extensive enough randomisation also allows a 'Tell Me' button to provide an explanation of the correct answer while retaining the usefulness of further attempts.

There are several ways in which the question in an exercise may be randomised. The context and wording of the question may be randomly selected from several alternatives. Questions involving data sets may use simulated data, perhaps generated from distributions with a variety of shapes and on different scales, or the data may be selected from a list of real data sets. Numerical values such as distribution parameters or other constants in the question may be randomly selected. Finally, an exercise may incorporate different variations, such as requesting the answer expressed as a proportion, a percentage or an odds ratio, and these variations can be randomised.

Independent randomisation of some aspects of an exercise does not work well. Consider an exercise about normal distributions with variations that ask for one of $P(X < a)$, $P(X > b)$, $P(c < X < d)$ and $P(X < e \text{ or } X > f)$. Independent random generation of a question type may present questions of the same type two or three times in a row, and some students would need to try an unreasonable number of questions before seeing all types. A workable solution to select one of n options is to constrain the randomisation by ensuring that there are no repeats in the most recent $(n - 1)$ choices and that all options occur in the most recent $(n + 1)$ choices. Separate randomisation of different aspects of a question should ensure that the same question is never repeated exactly (Stirling, 2008a).

Although this chapter emphasises the use of skill-based exercises for formative assessment, if the random generation of questions in an exercise includes a wide enough range of question types, the same exercise could potentially appear in a computer-based test for summative assessment, even after students have had unlimited practice with it.

19.5 Design principles for the CAST exercises

The remainder of this chapter discusses the design and implementation of a set of exercises for teaching introductory statistics (Stirling, 2008c) that is part of the CAST collection of e-books (Stirling, 2008b). These exercises are accessed using a web browser and, like the rest of CAST, are provided under a Creative Commons licence that permits free downloading and use.

The requirements of flexible question format, randomisation of questions and intelligent feedback necessitate implementation of the exercises with a high-level programming language. Java was used, owing to its widespread availability in web browsers. Each exercise is effectively a separate Java program (called an applet), allowing unlimited flexibility in question format and the potential for intelligent feedback.

The exercises were designed to follow, as much as possible, the following guidelines.

Working panels Panels containing interactive diagrams and formula templates should be provided to help answer the question, making each exercise self-contained without the need for pencil-and-paper or a calculator. An equally important function of these working panels is to provide feedback about the correct solution to the question (see below).

Feedback about wrong answers If possible, the exercise should identify the mistake in reasoning behind a wrong answer and explain this to the student.

Hints When a wrong answer is given, the exercise should give hints and allow further attempts. These hints may be supplied as textual messages or displayed in the working panels.

'Tell me' button Each exercise should include a button to show the correct answer by completing any working panels and explaining the solution textually.

Randomisation of questions Exercises should contain a button to randomly generate another version of the question by randomising as many aspects as possible, including the context.

The next sections illustrate these principles with a few CAST exercises.

19.6 Exercises requesting numerical answers

The exercise on the left of Figure 19.1 asks the student to guess the standard deviation of a data set 'by eye' from its stacked dot plot. A large error is permitted in this guess.

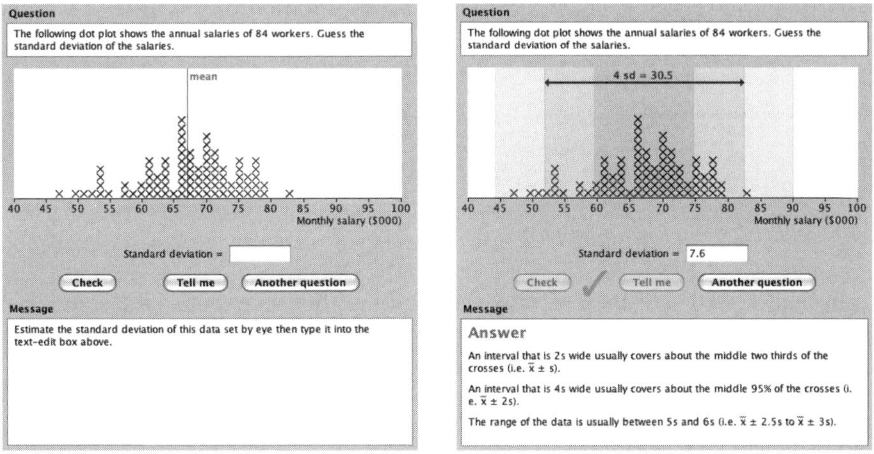

Figure 19.1 One numerical answer.

If the 'Check' button is clicked but the guess is too far from the correct standard deviation, the message box states whether it is too high or low, and advice is given about the proportion of values that should be within s and $2s$ of the $\bar{x}$. The diagram on the right of Figure 19.1 shows the result of clicking the 'Tell me' button; the bands within, $2s$ and $3s$ of the $\bar{x}$ are shaded to illustrate the answer.

The 'Another question' button randomly changes the context and wording for the question and the scale on the axis, and a new data set is simulated from a normal distribution with mean and standard deviation appropriate to the context.

The next two exercises form a sequence about the slope and intercept of a line. The exercise on the left of Figure 19.2 is a simple one in which the slope and intercept can be easily determined from the diagram. The exercise on its right is worded in terms of a least squares line. It does not display the predicted response at values 0 and 1 of the explanatory variable, so two templates are provided to help perform the calculations. If possible, it is always best to give a simple exercise as an introduction to a harder, more general one.

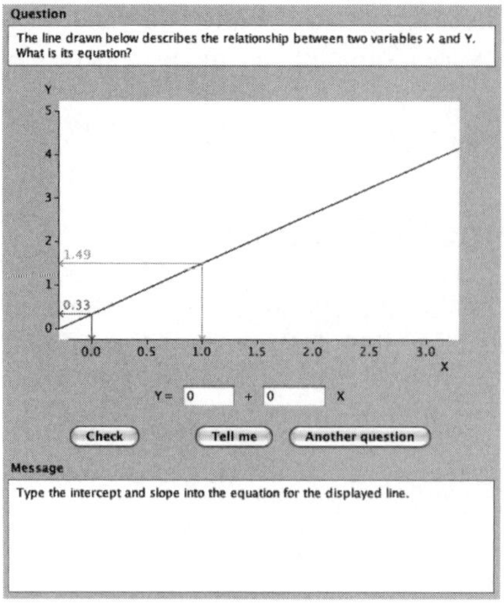

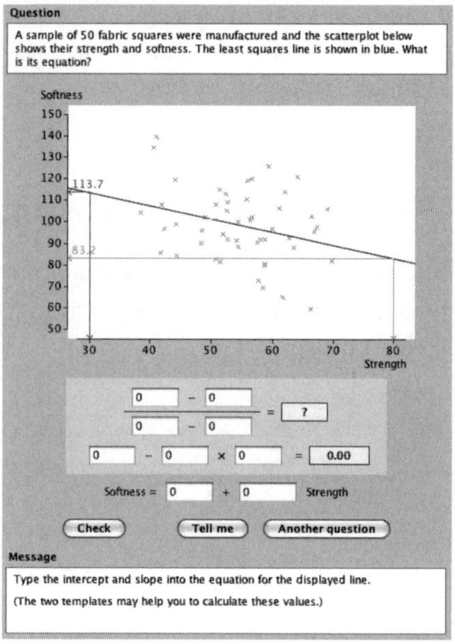

Figure 19.2 Two numerical answers.

In both versions of the exercise, the 'Tell me' button explains the solution in words in the message box. The exercise on the right also displays the calculations for the slope and intercept in its two templates.

The 'Another question' button changes the scales on the two axes of each exercise and the slope and intercept of the line. In the exercise on the right, the context is also changed and the data set underlying the least squares line is randomly generated.

19.7 Exercises requesting multiple-choice answers

In some types of exercise involving a numerical answer, it is unreasonable to expect an accurate answer; a choice between widely spaced alternatives is adequate. The exercise on the left of Figure 19.3 shows the pie chart of an ordinal categorical data set and asks for the combined percentage of values in several adjacent categories. The multiple-choice format is sufficient to assess whether students can read the information in a pie chart. Clicking 'Tell me' highlights the relevant pie chart area, and randomisation changes the context, data, targeted categories and multiple-choice values.

Multiple-choice is useful for questions involving alternative interpretations of information, such as in the exercise on the right of Figure 19.3. Randomisation changes the context, numerical values of the slope and intercept, the ordering of the two terms on the right of the equation and the ordering of the multiple-choice options.

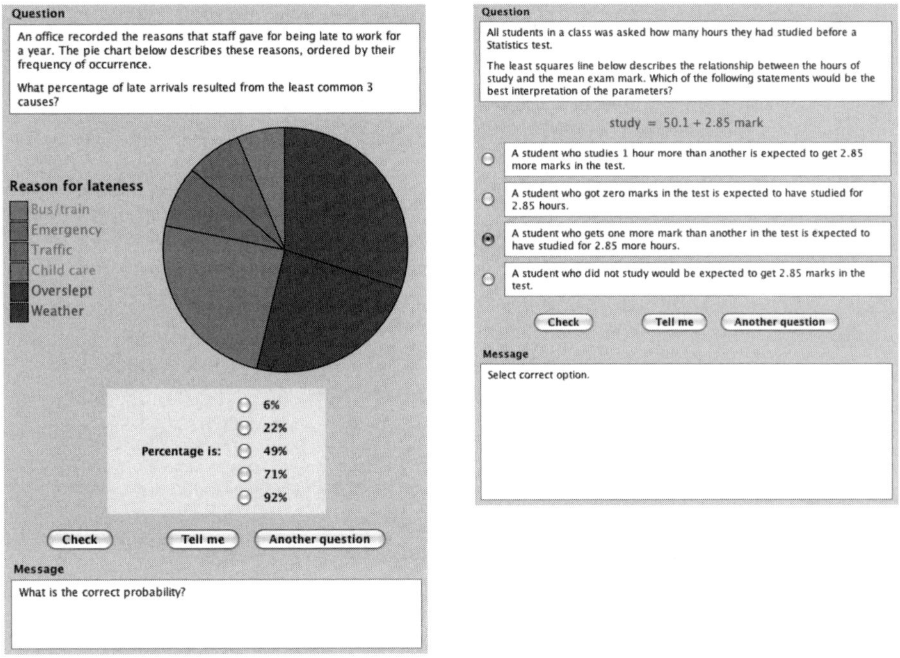

Figure 19.3 Multiple-choice answers.

The introductory exercise about the concepts of centre and spread that is shown in Figure 19.4 requires four multiple-choice selections, two with radio buttons and two with pop-up menus. Two distributions are displayed and the student is asked to describe the difference between their centres and spreads, both in terms of the question contexts and summarised by measures of centre and spread. The diagram shows feedback from a wrong answer, and randomisation involves the question context, the types of difference between the distributions and the ordering of the options.

19.8 Exercises involving interaction with a diagram

In many exercises the student must interact with a diagram, either as part of the working for the question or as an intrinsic part of the answer itself. The exercise shown in Figure 19.5 asks students to sketch a cumulative distribution function (cdf) from information provided in a histogram (note that the colours mentioned below relate to the online display of the exercises). The diagram on the left shows the feedback provided while dragging blue circles on the cdf at the class boundaries to adjust its shape. After clicking 'Check', the diagram on the right shows mistakes with red circles (shown here as darker grey) on the cdf and yellow shading in the histogram rectangles (shown here as lighter grey) where

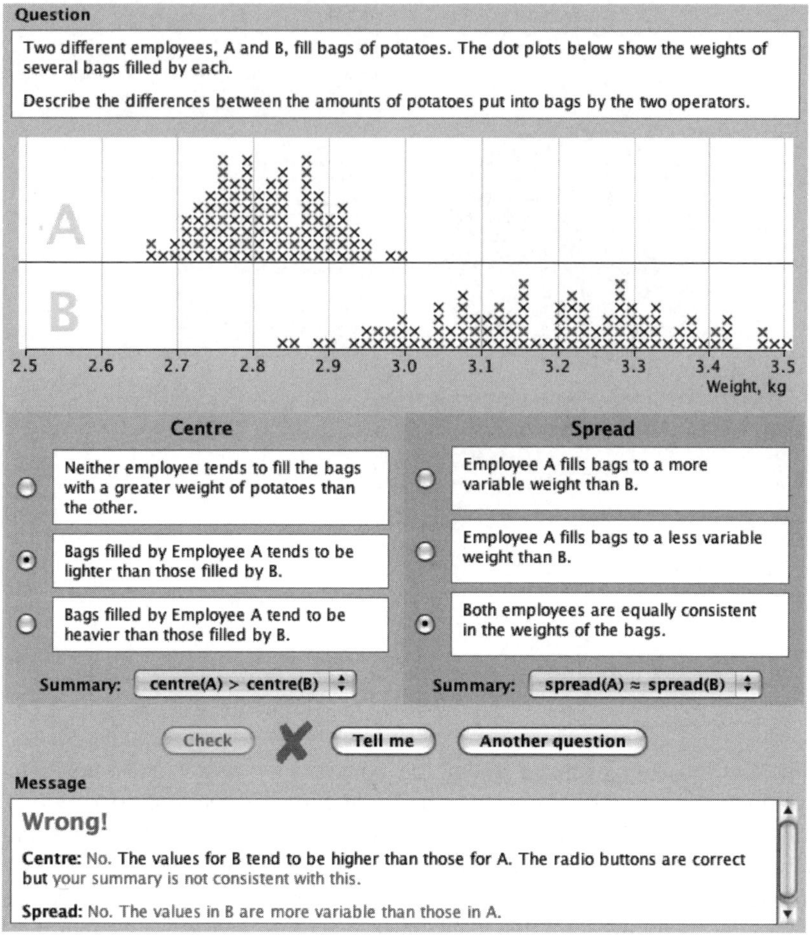

Figure 19.4 Several multiple-choice answers.

the cdf does not increase by the correct class proportion. Randomisation changes the context, the scale on the axis and the shape of the histogram.

The final two exercises are part of a sequence of increasing difficulty about normal distributions, in which interaction with a diagram is part of the working for the question. Both exercises ask similar types of question about a normal distribution. The context and normal parameters are randomised and the type of question is randomly selected from the forms

$$P(X < a),\ P(X > b),\ P(c < X < d) \quad \text{and} \quad P(X < e \ \text{ or } \ X > f).$$

The exercise on the left of Figure 19.6 shows three copies of the normal distribution. The two vertical lines can be dragged to read off areas under the

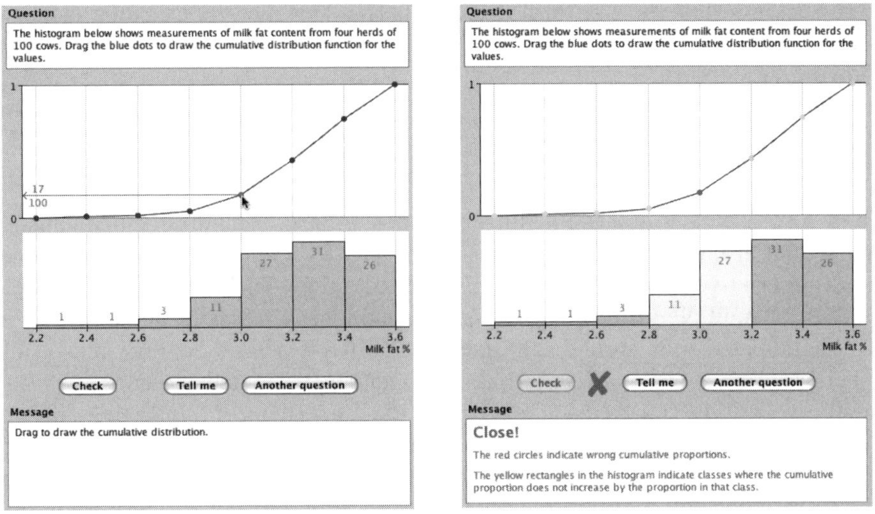

Figure 19.5 Drawing a diagram.

density function, or values can be typed in the two text-edit boxes to position them exactly.

The exercise on the right of Figure 19.6 replaces the distribution's density function with a standard normal density function and provides a template to translate the boundary values into z-scores. The diagram shows a typical question

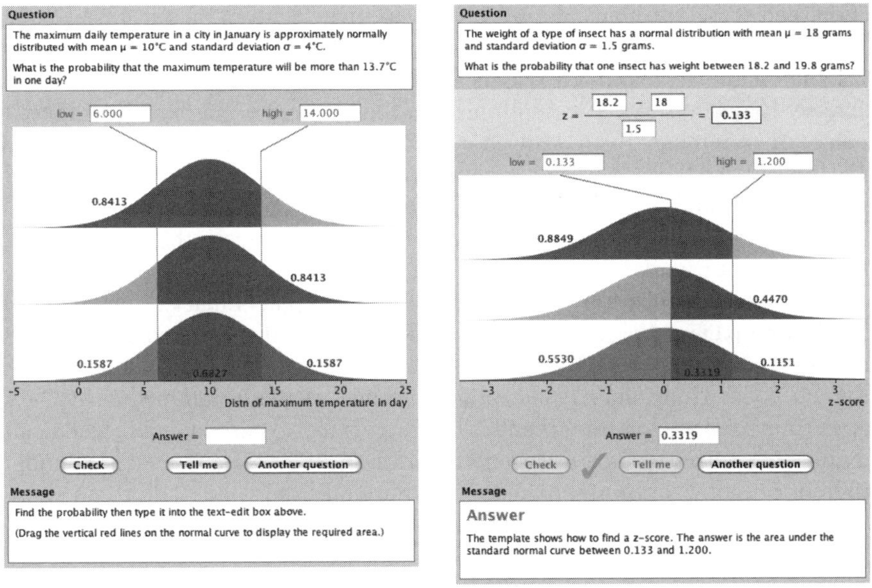

Figure 19.6 Finding values from diagrams and templates.

after clicking 'Tell me'. A third exercise in the sequence replaces the standard normal density function with a table of standard normal probabilities but is not shown here.

19.9 Implementation of the CAST exercises

The CAST exercises will eventually cover most topics in an introductory statistics course, but at the time of writing this paper only 91 exercises have been written about descriptive statistics and sampling distributions and there are none about models for multiple groups, regression or contingency tables or about inference. It is anticipated that around 200 exercises will be needed to give reasonable coverage of most topics in an introductory statistical methods course.

When writing hundreds of exercises, the importance of investing time in the design and development of a core library of common program code before writing any exercises cannot be overemphasised; common functionality will otherwise be repeatedly implemented for each successive exercise. The core library should include code to help lay out and display an exercise, support common actions such as clicking 'Tell me', randomise questions and generate random data sets, support common answer types such as multiple-choice questions, and draw simple graphical displays of data and distributions. An object-oriented programming language such as Java makes it relatively easy to encapsulate this common functionality in abstract classes from which individual exercises inherit it.

However, as illustrated in Sections 19.6 to 19.8, a collection of good exercises includes a wide range of different graphical displays, answer formats and interaction, and there are many different ways to give feedback about mistakes and to randomise the questions. These are often specific to a particular exercise and it is impossible to avoid a reasonable amount of programming for each new exercise. It is a mistake to compromise on the features of an exercise simply to reduce programming. Even with an extensive existing framework of Java code from other CAST e-books and a new framework of code for the exercises, it still takes the author between one and five days to implement a new exercise. Writing a large collection of good computer-based exercises therefore involves many months of work.

The CAST exercises are implemented as Java applets that are embedded in HTML web pages. Applets can be provided with textual parameters and these potentially allow aspects of the exercises to be changed without modifying their program code. Unfortunately the precise format of these parameters in CAST varies from exercise to exercise and is undocumented so, although it is technically possible for others to create variations on the CAST exercises such as adding different question wordings or data sets, currently only the author can do so. Documentation and tools to help others create new exercises are in the author's plans for the future.

19.10 Conclusion

Textbook exercises have always been an important learning tool, helping students to master the separate statistical concepts and tools that underpin more open-ended projects. Many of the limitations of static exercises can be overcome if they are presented and assessed on a computer. The computer can analyse a student's attempt and provide immediate customised feedback. It can also generate random versions of the exercise, allowing students to make repeated attempts at similar questions until the targeted concepts have been mastered. Finally, most students find that the interactive nature of computer-based exercises makes them more 'fun'. Design and development of a good set of exercises involves a lot of work but they provide an effective type of formative assessment that is worth the investment of time.

20

Assignments made in heaven? Computer-marked, individualised coursework in an introductory level statistics course

Vanessa Simonite and Ralph Targett

20.1 Introduction

While lecturers place great importance on tutorial exercises because of the contribution they make to students' learning, students appear to be more focused on assessment (Brown and Knight, 1994). In our statistics courses for non-specialist students, the timetable is designed so that a lecture introducing a topic in probability or a statistical technique is followed by a tutorial class designed to provide students with the opportunity to further their understanding by applying the material covered in the lecture to given problems or data sets. The tutorial class may also provide information on how to use statistical software and there will be opportunities for students to ask questions and to discuss problems or findings with the tutorial leader or with other students. In our experience, an increasing proportion of students do not attend tutorial classes, explaining that they plan to do the work at another time. In reality, however well intentioned, the

Assessment Methods in Statistical Education Edited by Penelope Bidgood, Neville Hunt and Flavia Jolliffe

majority of these students do not do the tutorial work and are disadvantaged as a result.

One way to ensure that students participate in the tutorial elements of a course is to include some of this routine tutorial work within the assessment, introducing a more obvious incentive for students to participate in these activities. The allocation of marks to tutorial work appears to be essential: Russell (2005) reported the impact of the introduction of weekly-assessed tutorial sheets in a course for first year engineering students. Student feedback showed an appreciation of the contribution that the weekly exercises made to learning but, despite appreciating their value, a high proportion of students acknowledged that they would not have completed the tutorial sheets if they had not been assessed.

While regularly assessed tutorial work may improve student engagement, the introduction of more assessment is likely to lead to a higher workload for lecturers in terms of marking. While traditional, non-assessed, tutorial work can legitimately be discussed freely between students, any work that is assessed must be carried out independently and is subject to university regulations on academic misconduct. When students are given the same data or set the same problems, academic misconduct can be a problem. In statistics and probability courses, the main difficulty is that students will copy each others' answers or work together (Hunt, 2005). A popular strategy for deterring plagiarism is to look for ways to individualise assignments, so that different students receive different data and therefore require different answers, thus deterring copying (for example, Simonite *et al.*, 1998; Jolliffe, 2003; Hunt, 2005, 2007a; Bidgood *et al.*, 2007). This means that creating an assignment is a more complex task. At Oxford Brookes University, the first use of automated, individualised coursework assignments was within a first-year introductory statistics module (Simonite *et al.*, 1998). Randomised parameter values were generated in Visual Basic, then the mail merge facility in Word was used to insert student names and the randomised parameter values into a coursework assignment. Different students were hence required to produce different answers. Solution sheets for each student were produced in the same way. By individualising the parameters in a problem, opportunities for copying were reduced. Hunt (2005) describes how this process can be simplified by using Excel rather than Visual Basic to generate the randomised parameter values used to individualise tasks and solutions.

While these approaches simplified the generation of individualised assignments and solutions, the task of marking students' work remained in the hands of the assessor. The drawback of setting individualised assignments was that although the individual solutions could be generated automatically, marking students' work was time-consuming as the model answers, varying from student to student, needed to be constantly referred to. The additional use of IT to automate the marking of assignments as well as to generate individualised questions and solutions was the next logical step.

20.2 Characteristics of an ideal system

The system developed at Oxford Brookes University embodies the following fundamental principles:

- Students will learn better if they carry out weekly tutorial tasks.
- Students will be more likely to comply if these tasks can be assessed.
- Assessments should be individualised to deter students from copying.
- Assessments should be designed to create minimal additional work for the assessor(s) and be marked and returned to students within a short space of time.

What, in these terms, is an ideal assessment system? Setting aside the content of the assignment, here we are concerned with the creation of a system for individualising assignments then delivering or distributing them to students, for collecting and marking students' work, for redistributing marked coursework and recording marks for administrative purposes. An ideal system would allow students to access their assignments and submit work independently of the assessor, and for assignments to 'mark and return themselves', and require assessors to exert minimal effort in managing the process. It would be secure, in that students should not be able to benefit by copying or by accessing the solutions to their version of a problem sheet without doing the work themselves.

Another important aim was to create a system that could be readily applied to other problem sheets, in other courses or modules and by other members of staff. Since creating a computerised assessment tool can be a major investment, we wanted to devise a system that would be as flexible as possible and could be easily reused on different assignments in statistics modules or in other modules setting work of a largely quantitative nature.

20.3 Simple mail merge assignment

To understand how the new system works, it is helpful to review the simpler process of using mail merge to generate individualised problem sheets. Designed to streamline and personalise mass mailings to lists of customers, the mail merge facility in Word places information from a linked spreadsheet listing individuals' names, addresses and other information into a standard letter, so that everyone on the mailing list receives a letter that has been personalised. In a statistics assessment the same idea can be used to create personalised assignments for each student. Instead of inserting the students' addresses and personal information, random generating functions in Excel are used to create 'personalised' parameter values or data for each student and these are inserted into the assignment. This means that while students will need to use the same methods, they will apply them to different values, and hence different students will require different results. Hunt (2005) describes this process in detail.

The first step in using this approach is to create a 'standard' coursework assignment in Word. Key elements, such as parameters or entries in a table, are represented by field names. A linked Excel file stores the names and student numbers of each enrolled student in the first two columns. In subsequent columns, the specific values of each field named in the assignment are entered using formulae that include random functions, enabling the parameter values to vary from one student to another. The mail merge facility in Word inserts each student's name, identity number and parameter values into the standard assignment, generating a unique, individualised assignment for each student. Individualised solution sheets are produced in the same way. Generating the problem sheets and solutions is relatively simple but a drawback is that although the solutions are produced automatically, marking is time-consuming: the model answers need to be constantly referred to as the answers vary from student to student and hence cannot be memorised.

While combining individualised coursework with more frequent assessment appears to address both the desire to deter plagiarism and the need to increase student participation in scheduled tutorial work, a consequence is a potentially large increase in the workload for the assessor. This chapter describes a system that was designed to achieve more frequent, academically 'secure' assessments which would, as far as possible, 'mark themselves'.

20.4 A new system

A new system was tested in one of our introductory statistics modules, Basic Survey Methods, which introduces first-year undergraduate students studying mathematics, statistics or marketing to statistical methods used in the design, analysis, reporting and evaluation of surveys. The module is applied in nature and regular attendance and participation in tutorial work is essential. The module is taken by approximately 80 students per year, who attend lectures together and are divided into sets of approximately 20 students for tutorial sessions.

Within the new system, these students access individualised problem sheets, submit their answers and receive their marks and feedback via a password-protected web-page. The distribution of assignments to individual students, the submission and collection of their work and the distribution of marks and feedback is handled online. Within a web-based system, these activities can take place independently of the assessor and wherever the student can access the Internet.

The assessor writes the assignment and specifies the formulae for generating the parameter values and the solutions, then runs one program to automatically individualise assignments and another to mark students' work. The programs are written in Visual Basic (VB). The 'CALCULATEMERGE' VB program generates the individual assignments and the 'MARKING' VB program marks students' work automatically, providing feedback for students and a mark sheet for the lecturer. These two programs were designed to be independent of the content of assignment itself, as will be seen below.

The advantage of separating the generating and marking activities from the specification of questions and answers is that the programs can be used again and again, without modification. This provides a valuable return on the initial investment of effort required to create the programs. The next section describes how the system works from the students' and assessor's viewpoints, and shows the assessor's inputs for a simple problem sheet consisting of questions on standard errors and sample sizes.

20.5 Setting up an assignment

To set up an assignment, the assessor must create a standard problem sheet, a list of students and a spreadsheet. The problem sheet is in the form of an HTML web page, showing the questions as they are to be displayed to students, with the parameters to be individualised represented by field names enclosed within '&' signs. Figure 20.1 shows a problem sheet as it appears before a student has entered their student number and password. At the top of the page, a message shows how many answers are required. Question 1 is an exercise in calculating standard errors and sample sizes. The question refers to two surveys reporting the percentage of respondents who buy more than 50% of their everyday shopping from one store. Students are asked to calculate the standard errors of these percentages. Figure 20.1 shows the field names that will be replaced by individualised values, namely the sample sizes for areas A and B, denoted by &nA& and &nB&. Once a student enters their student number and password, field names are replaced by individualised values, so that each student is working with different sample sizes and percentages. The problem sheet can be printed and students can access it as many times as they need. Each time they access it, the same parameter values appear in response to their unique student number and password.

The second element that an assessor needs to set up consists of a list of students enrolled on the module, including names and student identity numbers. This is simply downloaded in CSV format from the university's student management computer system.

The third element is a spreadsheet specifying, for each question in the problem sheet, the formulae for generating individual values for each parameter, the formulae for calculating the answers, the accuracy required of each answer and the marks available. In the example below, the answers required are shown in numeric form. Figure 20.2 shows the spreadsheet displaying the values for one student for the questions seen in Figure 20.1.

The positions of some elements of the spreadsheet are fixed: the cells in row 16 are fixed and the questions always begin in row 18. These fixed points mean that the VB CALCULATEMERGE program can always find where the

8400 SAMPLING AND SE'S PARTICIPATION EXERCISE

Your answers will be stored on the server, only one version at any one time.

There are a total of 5 answers required:

Questions shown in red must be answered.

Input your 8 digit student number:

Input your password for this module:

Enter your name:

Work out the answers to the questions below and then write your answers into the boxes below:

1. Customer loyalty surveys are carried out to determine the percentage of people who busy more than half of their everyday shopping from the same store. In two areas, the results achived in surveys asking this question were as shown below.

(1a) Calculate the standard error for the percentage of people who buy more than half of their everyday shopping from the same store, for each area.

Area	Sample size	% Who buy more than 50% from one store.	Enter answers
A	&nA&	&A&	se A
B	&nB&	&B&	se B

(1b) In area A, how large a sample size would be needed to ensure a standard error of 1.5%:

Figure 20.1 Sample problem sheet in HTML, showing field names.

parameter values begin. The most important section of the spreadsheet starts in row 18, column C where each question or part question that requires an answer is listed. In this example, Question 1 requires three answers, two associated with 1a and one with 1b. Question 2 requires two answers: one for 2a and another for 2b. To the right of column C, there are three pairs of columns in which the parameters for each question are named and values given. For example, column D shows

	A	B	C	D	E	F	G	H	I	J	K	L	M	N	O	P
1	**8004 Sampling & SE's Answer sheet**															
2																
3	**Note: Always SAVE with cursor over Student#**															
4																
5	**Note: All decimal numbers to 3 significant digits (for marking)**												Answer Format			
6			Decimal marking is done to a set tolerance between D1=0.1 through D9=0.9										D = decimal number			
7			Example: D5, If answer is 101.5 then 101 and 102 are both accepted										I = Integer up to 32000, else Text			
8													T = plain text max 256 characters			
9				Algorithm to avoid the the display of 0.0276 as 2.76E-02 in merge fields												
10				sereqd	0.02773											
11				3 figures	277											
12				text	0.0277											
13																
14																
15	**Note: Always SAVE with cursor over Student#**															
16			Question	Student #												
17				Merg Field	Merg Val	Merg Field	Merg Val	Merg Field	Merg Val						seA	Mark
18			1a	nA	1097	A	34						ANSwer	D3	1.43	1
19															seB	
20			1a	nB	509	B	52						ANSwer	D2	2.21	1
21															samplec	
22			1b	area	A	percent	1.50						ANSwer	I	997	1
23															se	
24			2a	ssizeq2	58	mean	2.24	sd	0.210				ANSwer	D5	0.028	1
25															sizereq	
26			2b	sereqd	0.0217								ANSwer	I	94	1

Figure 20.2 Spreadsheet.

the two sample sizes nA and nB, mentioned in Question 1a and to the right, their values for a particular student. For each question, the spreadsheet shows the field name associated with each parameter and its (individualised) value, supplied by a formula, on the right. In order to individualise the assignment, each of the formulae for the parameter values includes at least one random element. The number of parameters listed alongside each question may vary: here, there are questions with one, two or three parameters listed on the same line.

In column M, the word ANSwer marks the start of the solutions. Three items of information are shown: column N shows a code representing the level of accuracy required for a correct answer. This allows students whose answer lies within a certain distance of the correct answer to be marked as correct. A more sophisticated system might allocate partial marks for answers that were close but perhaps not close enough. Column O shows the correct answer and column P shows the number of marks available for a correct answer. Brief feedback statements can be entered in column Q if required.

In the example shown here, the largest number of parameters needed for any question is three, for question 2a. To cope with a maximum of three parameters, three pairs of columns, D and E, F and G and H and I, were assigned to the field names and formulae for the parameter values and hence the ANSwer column is in column M. If more parameters are required, then further pairs of columns will be devoted to their field names and formulae, and the ANSwer column and those to its right, shifted to the right accordingly. Additional questions are simply added as additional rows. It is convenient to arrange the parameters in line with the questions that they first appear in, but the answers can draw on parameters from any question. For example, in a question with several parts, there may be a key parameter, perhaps a sample size that is used to determine a number of answers. This parameter would only need to appear once on the left of the spreadsheet even though it was used in a number of the answers listed to the right of the spreadsheet.

20.6 Files for each student

The system operates by creating up to four small CSV files for each student. Four files are created for students who complete the assignment, two for students who are registered, but do not complete the assignment. File names indicate the content or purpose of the file and are indexed by student numbers. A student with student identity number 123 who completes the file will generate four files. Table 20.1 lists the content of each type of file and the stage at which they are generated.

Table 20.1 Student files.

File name	Content	Created by
Q123	Individual parameter values to be inserted into questions	CALCULATEMERGE program
A123	Correct answers	CALCULATEMERGE program
S123	Answers submitted by student	Student
M123	Marked answers and feedback	MARKING program

20.7 Generating the individualised problem sheets

After creating the problem sheet, student list and spreadsheet, the assessor runs the CALCULATEMERGE program. This program is stored on the assessor's computer. Once the program is open, the assessor simply has to input the location of both the CSV file listing the students and the spreadsheet. The CALCULATEMERGE program reads the student list and uses the spreadsheet to generate two sets of CSV files: the first set, the Q files, contain the individualised parameter values for each student and the second set, the A files, contain the answers for each student based on their individual parameter values. For student number 123, for example, these are the file Q123, containing the individualised values of 10 parameters, and the file A123, containing the corresponding five answers. The assessor copies the files of parameter values (Q files) onto the web server from which students will access their individual assignment. As these files contain only the parameter values and not the questions, they require little space. The files containing the answers (A files) are not transferred to the web server.

20.8 Students' inputs

A link is provided from the opening web page to a specialised web page that allows students to obtain a password, which is sent to them automatically by email. Another link allows the student to use their password to see the coursework assignment, as shown in Figure 20.1, but with field names replaced by individualised parameter values. As mentioned above, the assignment can be printed off and will show the same parameter values whenever it is accessed by the same student. As students submit their answers via the web page, the third set, of 'S' files, is generated, storing the answers submitted by each student.

20.9 Marking students' work

At the appropriate time the assessor copies the S files, storing the answers submitted by students, from the web server to the assessor's computer and runs the MARKING program. The assessor's role is to identify the locations of the list of students enrolled, the S files storing students' answers and the A files storing the correct answers. The program then goes through each pair of A and S files, comparing the submitted answers and the correct ones, awarding marks accordingly. Output from the marking program consists of a marked, M file for each student and an updated student list for the assessor, showing the marks achieved by each student. The M file lists the correct answers alongside those submitted by the student, with brief feedback and the marks awarded shown next to each item. The assessor then copies the files to the web server so that students can access them. Access to each set of student marks is password-protected.

20.10 Practicality

In creating the system, the key aims were to produce a method that would allow students' tutorial work to be assessed regularly without creating an excessive workload for the assessor, with fewer opportunities for academic misconduct than would be offered by a conventional problem sheet.

Initial trials were carried out within a basic survey methods module, testing each student's ability to calculate standard errors, confidence intervals and required samples sizes. These showed that the system functioned in practical terms: students were able to access their assignments and submit answers and the marking program supplied feedback and a mark sheet as intended. The system has now been extended to other assignments and has also been used successfully in a large computing module, to assess a part of the module involving mathematics.

20.11 Security

Two aspects of security that need to be considered are: the extent to which materials held on the server are open to interference, for example, from hackers; and the extent to which the assignment is vulnerable to academic misconduct such as copying.

Students' passwords are not stored on the web server, only an encrypted hash of the password. This way the password cannot be read or hacked. Other files on the server are: the assignment (HTML web page); student parameter files; student submitted answer files; and student feedback. The spreadsheet containing answers and the individualised correct answer files remain on the assessor's computer. Since correct answers and formulae for calculating them are not stored on the server, the potential for students to access solutions dishonestly is limited.

Since different students require different answers, dishonest students have nothing to gain by copying. Academic misconduct in the form of two or more students working together will at least require the participants to do the assignment more than once, so that additional work is required. For some students this may be a sufficient deterrent. Impersonation, in which someone other than the student does their work for them, remains a risk unless the assignment is carried out under test conditions with identity checks.

20.12 Effort required from the assessor

The time invested in setting up the system for the first time was divided between creating the programs that generate and mark individual assignments and constructing the files containing the assignment in HTML and the spreadsheet specifying parameters and solutions. Now that the programs have been developed, the assessor only needs to create the HTML file of the assignment and a new spreadsheet of parameters and answers. The workload for the assessor in terms of individualising tasks and marking is also very small, being made up of the time taken to copy files from the assessor's computer to a server, and vice versa, and to input relevant file locations to the CALCULATEMERGE and MARKING programs.

20.13 Flexibility

In terms of flexibility, a successful approach to individualised and automatically marked assignments should allow as much freedom as possible for the assessor to choose the environment in which students complete the assignment and the type of content in the assignment and to modify or introduce new assignments as and when they wish.

In terms of the environment in which students work on the assignment, assessment could be carried out under test conditions, during a scheduled tutorial, or as a 'take away' assignment. All of these are possible with the system described here and have been used in practice by different members of staff.

In terms of the type of assignment, in the example shown above, all the problems required numerical answers, but the system has also been applied to a problem sheet incorporating multiple-choice questions designed to check students' interpretation of hypothesis tests and other conclusions.

The role and construction of the spreadsheet is a design feature that enables the system to be readily applied to a new set of questions. The CALCULATEMERGE generation program, is able to use the spreadsheet supplied by the assessor to calculate and individualise the information for each student, no matter how the questions, parameters and answers may change from one assignment to another.

As explained earlier, the generating and marking programs can be applied to any new assignment without amendments, enabling the easy transfer of the system to a new assignment. In other words, what we have is a *system*, rather than just an example, for generating and marking individualised assignments. As a consequence, the investment of effort needed to create new assignments is small and this means that the system can be reused very efficiently even in modules with relatively small class sizes.

20.14 Conclusion

The most important advantage of the system described above is the flexibility which has been designed into it. The key feature is that the part of the system that was most labour-intensive to develop, namely the VB programs that generate and mark individual assignments, being independent of the assignments themselves can be applied to new assignments or a new course without further programming effort.

21

Individualised assignments on modelling car prices using data from the Internet

Houshang Mashhoudy

21.1 Introduction

In this chapter, we introduce four assessment tasks based on new and used car data. It is shown that such data, collected carefully, can be used for individualised assessments of both simple and more complex statistical techniques.

The idea of setting assignments in statistics that are based on real data is not new (Trumbo, 2002). Increasingly, teachers of statistics recognise the pedagogic advantages of students collecting their own data (Obremski, 2008). There are some published articles on the use of real data sets for regression assignments. For example, McLaren and McLaren (2003) suggest a series of assignment questions based on regression and forecasting methods applied to electricity bill data they had collected. Pardoe (2008) considers the application of advanced regression methods to modelling house prices based on a data set he had collected. Although these articles can be used as a basis for assignments, they omit instructions on data collection and information on the data sources.

The car market is a familiar and easily understood context with which students can engage, since many of them will have experienced buying cars either personally or with their families. The market for used and new cars is a huge industry and it is easy to convince students of the usefulness of completing an assignment in this area. The context is particularly relevant to students on

Assessment Methods in Statistical Education Edited by Penelope Bidgood, Neville Hunt and Flavia Jolliffe

business or economics courses but, because of its general and topical nature, it can also be used with students on other courses.

Examples can be found in textbooks, based on relating the price of used cars to their age or mileage. More recently Obremski (2008) suggested the collection and analysis of such data as a practical class exercise. There are also some research articles, mainly appearing in economics journals, that deal with specific aspects of the car market (Engers *et al.*, 2009; Betts, 2006). Car prices have been studied across many different countries; for example Kooreman (2006) identified some anomalies in used car prices in the Netherlands, such as a sharp price reduction once a car exceeds 100,000 km. Matt (2008) modelled the asking prices of used cars in the USA by age, mileage, make and condition and, most unusually, whether owned by smokers or non-smokers.

In this chapter the emphasis is on the complexity of the data collection process, the study design and the need for clear guidelines and instructions. Relating car prices to other variables is not as straightforward as suggested by many textbook examples. Four different scenarios are introduced, together with the relevant data sources and details of the data collection. Several possible tasks are suggested and it is left to instructors to select those most appropriate for their students, who could work independently or in groups to collect data. The scenarios include variables which might lead to fairly strong relationships between them, and can be used to develop several possible models.

21.2 Individualised data and plagiarism

Plagiarism can be a problem and, with the advent of social networking web sites, students may even be tempted to obtain copies of work completed in previous years. The use of individualised assignments can help ease this problem (Bidgood *et al.*, 2007). For the assignments introduced in this chapter students collect their own data using sources that change on a daily basis. This way of individualising the data is very effective in tackling plagiarism.

Assignments based on individualised data do take longer to mark. To reduce the marking load, instructors could design specific questions, give clear instructions, introduce group work, or design tasks that require using Excel or statistical software. Some examples are given later in this chapter.

Some popular UK web sites are introduced here. Similar web sites from other countries are also available and easily found using Internet search engines. It should be straightforward to adapt the instructions for data collection for use with other web sites.

21.3 Outline of the assignments

Four distinct assignment scenarios, each with a set of specific tasks, are introduced here. Each assignment consists of two stages: the data collection stage and the data analysis stage. The assignments also require some basic knowledge of the context of study before data collection and analysis; this is described in Sections 21.4 to 21.8. The context could be blended in with the assignments or provided as a separate document.

Although these assignments are different in nature, the methods of analysis required are similar, so every year a different one of these can be set for the same module. Also they can be set for several different modules in the same year. Only one of these assignments should be used with the same group of students. Different versions of these assignments have been used by the author with students on business courses as well as mathematical sciences courses for well over twenty years. Some versions have been used for group assignments. Group work is particularly useful for collecting and compiling larger data sets.

Data collection tasks always generate much discussion amongst students, as there are many points that need careful consideration and clarification when collecting data on cars. With some classes, the data collection tasks were made a class activity so that the students could interact with each other and get feedback from the instructor.

Initially male students are generally more enthusiastic, but female students also appreciate the relevance of the car market and soon realise that the technical aspects, such as the engine size and the brake horsepower only require common sense, and they usually end up producing some of the best reports.

Briefly, the four assignments are:

1. Modelling the prices of new cars based on their weight, 'brake horsepower', global origin and engine size.
2. Modelling the asking price of a used car using mainly its age and mileage, also considering some other factors.
3. Modelling the asking prices of different cars of a certain age.
4. Modelling the actual price of a used car obtained at an auction.

For all the assignments, students are required to report on data collection, with a discussion of any shortcomings of the data collected, providing details of data sources and a table of their data. They enter the appropriate data into a suitable package and carry out a full regression analysis that includes residual analysis and prediction. Scatter plots can be used to show group differences and possible interactions. In a written report students describe and interpret the results of their analysis and the meaning of the coefficients in context, discuss models fitted and their appropriateness, their advantages and disadvantages, and select the best models. Possible more advanced tasks include investigation of interactions, quadratic terms or log transformations.

Later in this chapter, a list of tasks that are specific to each scenario and further details for instructors are provided. These can easily be modified to reflect different approaches and the level of teaching.

Any marking scheme should reflect the tasks set, the level of work expected, and how long each task is likely to take. It is suggested here that for the first and the fourth assignment 10% of the total mark should be allocated to data collection but for the other two assignments this should be 20%.

It is advisable to check the students' data before they start their analysis to ensure they have collected appropriate, individualised data. It is all too easy for students to collect worthless data if they do not have clear guidelines on the study design.

21.4 Data collection issues

The major shortcoming of most textbook exercises on car market data is their over-simplification by ignoring many factors involved. These issues are discussed here before the specific assignments are introduced.

21.4.1 Make of car

Some manufactures can expect higher prices because of their brand image or prestige or because they are known to produce better quality cars. Prices of cars could also depend on the global origin of the manufacturer, for example German cars tend to be more expensive than French ones. The effect of global origin has been widely studied in the marketing literature (Haubl, 1996). To avoid the problem that some manufacturers have factories in different countries, global origin is assumed here to mean the country of origin of the ownership of the car company and not where the cars are made. Table 21.1 shows examples of the global origin of cars selected by our students.

Table 21.1 Examples of global origin of cars.

Global origin	Make of cars
France	Peugeot, Renault, Citroen
Germany	BMW, Audi, Volkswagen, Mercedes Benz, Porsche
Japan	Toyota, Nissan, Honda, Mitsubishi, Subaru, Suzuki, Daihatsu

21.4.2 Model of car

Car manufacturers often make several models, ranging from small to large – for example, Ford makes Fiesta, Focus, Mondeo, etc. The features of such models as sports cars, four wheel drives and people carriers will have a significant effect on the price, and so it is important to fix certain conditions in the data collection in order to make the data analysis more manageable. It may be helpful to exclude sports cars, for example, since their inclusion could result in many unusual observations. It would be possible to introduce a dummy variable to distinguish sports cars, but for a small assignment it is best not to have too many categorical variables.

21.4.3 Editions

Car manufacturers produce several different sub-models, often referred to as editions or derivatives. Different editions of the same model may have different engine sizes, trim levels, body types (e.g. saloon, hatchback or estate), number of doors, fuel types, or types of transmission. Editions are described and distinguished by use of certain terms or abbreviations, for example TD means Turbo Diesel. It is useful, but not essential, for students to become familiar with these for the cars they choose. For example, Ford makes a model called Focus which has several editions, including the 1.4 CL Manual 5 Door Hatchback Petrol and the 1.8 TDi Ghia Manual 5 Door Estate Diesel. Most models of the same car will have a cheaper, base edition, but also different editions of the same car may vary significantly in price, as those that are better equipped or have a higher performance have a higher price. Some editions are distinctively different, such as those with diesel fuel, sports editions, cabriolet (open top) and automatics.

21.4.4 Age

Car advertisements usually only indicate the year of registration for a used car. In the UK, the registration plates for cars manufactured and registered after 2001 can be used to approximate the age of the car. The age identifier (i.e. the date of registration) appears on the registration plate as the number in between the two groups of letters (e.g. AB 06 CDE, where 06 indicates the car was registered between 1 March 2006 and 31 August 2006). If the number 56 appeared instead, then the car would have been registered between 1 September 2006 and the end of February 2007. Hence the date of registration can be approximated as either 1 June or 1 December, the mid-points of the above two periods. Similar numbering schemes operate in other countries. The age of the car can thus be approximated and recorded in months. The exact month of registration could be obtained by contacting the dealers, but this is time-consuming and so not recommended when a large number of students are doing the assignment. For a new car, the age should be recorded as zero.

21.4.5 Mileage

It makes subsequent analysis easier if mileage is recorded in thousands of miles (or kilometres). When interpreting a regression coefficient for mileage, what is of interest is how much the price of a car changes for every 1000 miles travelled. Since there is typically a high correlation between age and mileage, it is found that mileage per year is often a better variable to use. Large differences in mileage of cars of the same age would make a difference to their asking prices. Hence, it is better to collect data on high and low mileage cars for all ages, defining, for example, high mileage cars as those above 15,000 miles per year and low mileage cars as those below 10,000 miles per year. Mileage could then be included as a categorical variable.

21.4.6 Dealerships

There are price differentials between different market outlets. Asking prices advertised by main dealerships are amongst the highest because their prices tend to include longer warranty, generous part exchange allowances and a service. Their prices are also subject to negotiation, leading to special discounts. Car supermarket prices are generally lower because they are advertised for quick sale. Main dealerships and car supermarkets tend to sell cars that are under three or four years old. Non-franchised dealers tend to sell cars of a more varied age range and their prices are somewhere in between the other two types of dealer.

21.4.7 Other factors

There are many other factors that affect the price of a car and it is generally not possible to model the effect of all these simultaneously. Examples of such factors include colour (with metallic colours being more expensive), fuel consumption, insurance group, air-conditioning, inclusion of an airbag and central locking. Many of these are closely related to the model and edition of cars, although some could be used as classification variables. For example, a possible task might be to compare prices of cars with high and low fuel consumption. Used car prices can also be affected by service history, number of previous owners, and months remaining of tax or MOT (British roadworthiness test). For the assignments mentioned here some of these factors will balance each other out and the models fitted using the main variables tend to determine a large proportion of variability in prices.

21.5 Assignment 1: Modelling the prices of new cars

The main object of this assignment is to construct models for predicting the price of new cars from their (kerb) weight, brake horsepower, global origin and engine size. Analysis of such data is not new, (Lock, 1993). The problem with such studies is that the data used is taken at one point in time, soon becoming out of date. This assignment overcomes such problems by making use of the latest available data from a popular United Kingdom web site that is regularly updated.

21.5.1 Instructions for data collection

- Choose cars from two or three different global origins.
- The models of cars chosen should be from a wide range of weights: light cars as well as heavy cars. Otherwise the regression model will not be so useful for predicting price from weight.
- Choose only one edition from each model because the weights of different editions of the same model will be very similar, all having the same shape.

It is particularly useful to discuss these points in class. For example, to be consistent the instructors could suggest the use of the base edition, or one of the cheapest in the range.

It is possible to use the car manufacturers' web sites to get the above data but experience suggests that this is very time-consuming. In the United Kingdom, data can be obtained more easily from the *What Car* web site (www.whatcar.com). Students are given detailed instructions on obtaining data from this site and advised to collect a sample of at least ten cars for each of two or three global origins from a whole range of prices and sizes. For the first assignment, students require data on weight, engine size, brake horsepower, make, model, the edition of the car with details of its specification and, finally, the target price which is the best new price that can be obtained in the market.

Recording the data from the web site is not as straightforward as one might imagine. For example, the engine size of a car may be presented as 1498/4, indicating an engine capacity of 1498 cc and four cylinders. The number of cylinders should in fact be ignored but, in the absence of any advice, a student could be forgiven for thinking the engine size was 374.5.

21.5.2 Some specific tasks

- Fit various regression models to predict the new car price from variables: engine size, weight, brake horsepower, and global origin.
- Carry out a residual analysis to identify cars that are better value, or the opposite.
- Instead of using global origin as a categorical variable, introduce a dummy variable to represent make of the car. For this task it would be essential to choose makes with many models or there will not be enough cars for comparison. Alternatively, compare a group of popular makes with a group of prestige cars.
- A simple task would be to calculate the price per kg for each car and give appropriate summary statistics for this. It is also possible to analyse this variable by global origin or use it as the dependent variable in a regression model.
- Repeat the analysis, replacing weight with a categorical variable that describes the car size (i.e. small, medium and large).

21.6 Assignment 2: Modelling the asking prices of used cars

This assignment involves the use of multiple regression analysis to predict the asking price of a certain used car from its age and mileage. It could also allow for other variables such as the edition of the car.

21.6.1 Choosing a car

Students must first select two editions of cars that are very different in price from a specific make and model. It is preferable to ask students to choose two editions that are very different in price. For example, a bottom of the range Ford Focus such as the 1.4 CL Manual 5 Door Hatchback Petrol would be much cheaper than a top of range Ford Focus of the same age and mileage, such as the 1.8 TDi Ghia Manual 5 Door Estate Diesel. The car should have been manufactured for some years to ensure there are enough used cars available. Students need to collect data on cars of various ages and so it is suggested that they have at least one car from each relevant registration period. It is not a good idea to include cars that are older than seven years because their price is likely to be affected by their condition and other attributes.

21.6.2 The sample and the data sources

United Kingdom Internet sources that carry real-time classified advertisements for used cars include www.autotrader.co.uk, http://www.exchangeandmart.co.uk/motoring and http://www.parkers.co.uk/carsforsale. These can easily be used to obtain a sample of at least 10 for each of the two editions of the car chosen. It should be noted that these sites require a postal code in order to initiate a search. It is possible to put a distance limit on the car's location from a given post code. A nationwide search is also possible, by increasing the distance limit and so increasing the number of cars found. Most cars on these sites are advertised by main dealerships and students should search these only and avoid private sales. It is unlikely that small differences in the engine size or trim would affect the used prices of different editions. It is therefore possible to ask students to model prices of a group of editions that are very similar perhaps using new prices as a guide. This is particularly useful when students choose less popular cars that may have fewer numbers on sale. There are similar, relevant web sites in other countries, for example, www.cars.com in the United States.

21.6.3 Some specific tasks

- Calculate mileage per year and summarise your results.
- Produce a labelled scatter plot of price against age where each point is labelled with the mileage per year.
- Fit various regression models to predict the asking price from age, mileage and edition. Also consider the use of mileage per year instead of mileage.
- Estimate the asking price of a nearly new secondhand car with delivery mileage only. (This would be the constant term and students need to appreciate that it is an extrapolation with its usual faults. This could also be

interpreted as how much a new car is worth in the used market immediately after it is bought, proving a useful piece of information to those who intend to buy a brand new car.)

- Fit a log linear model that includes age only. Give the plot of data with the fitted curve superimposed. Use the coefficient of age, 'β', say, to estimate the percentage difference in prices of cars of different ages using 100 [exp (β) − 1]. To estimate the year on year depreciation using 100 [exp(β) − 1].
- Comment on outliers. Large residuals can be interpreted in this context as cars that are different in their condition or some other attribute such as having metallic paint or being an ex-police car.

21.6.4 Tasks for larger assignments

- Include region as a categorical variable. Appropriate postcodes could be used to conduct a search in two or three different regions.
- Introduce a categorical variable to compare prices from different dealer types. The data could then be directly collected from the dealers' web sites.
- Mileage can be used as a categorical variable, as explained above.

21.7 Assignment 3: Modelling prices of different cars of a certain age

This assignment is similar in most aspects to the first, except that the price of a particular age of car is modelled, rather than the new price. Here the most important independent variable that determines the used price is the current new price of the car but other variables such as weight and make, brake horsepower, mileage and engine size could also be considered. In the UK, different registrations can be used for different student groups in the same year. For example, the 06 registered cars could be given to tutorial group A and 07 registered cars to tutorial group B.

21.7.1 Instructions for data collection

From the used car web sites mentioned above, students collect data on the asking price of about 20 or more different models from a given registration, across the whole range of prices and sizes. Students should use the exact make, model and edition of their cars to get the current new price, the weight and the other measurements for their cars using the *What Car* web site. The issues raised in previous scenarios are also relevant here and should be given out as guidelines: for example, collecting just one edition from each model of car if weight is to be investigated as an independent variable.

21.7.2 Some specific tasks

Most of the tasks discussed for the first assignment and some of those for the second are also relevant here, for example, use of mileage as an independent variable. In addition:

- Students could fit regression models to predict the used price from variables current new price, weight and mileage etc.
- Mileage could be considered as a categorical variable (see Section 21.4.5).
- If log of used price is modelled with the log of the new price as a variable in the model then the regression coefficient for this variable can be interpreted as a partial elasticity showing the average percentage increase in the price of a used car for a one percentage increase in the price of the new car.
- Use global origin or make as possible categorical variables.

21.8 Assignment 4: Modelling the actual selling price of a used car

It is possible, and perhaps more sensible, to model prices from car auctions, because they record the actual sale price of the car rather than the asking price. A statistical model would then give buyers an idea of the lowest price they should pay for a particular car. The main problem here is that currently there is a fee to access data from most car auction sites. Paying a fee may be justified for a large-scale project carried out by one student but not for a class assignment. However, www.centralcarauctions.com provides a useful free resource in the UK. The data can be accessed by selecting the 'Price guide' tab from the home page of this website.

The data on this site are organised in such a way that data collection is very easy. The tasks mentioned for Assignments 2 and 3 can also be used here. A possible problem is that in future this web site may not be accessible free of charge, although others might start providing a similar service. It is advisable to carry out an Internet search on the current policy of such web sites regarding the provision of free data to the public.

21.9 Conclusion

The assignments discussed in this chapter serve many purposes, such as the collection and use of real data from the Internet, individualised assessment, group assessment and problem-based learning. The use of up-to-date data of general interest motivates students whether they are in specialist or service courses.

Data collection forms a part of the assessment here. Web sites change, so checks need to be made to see if instructions should be altered. Some models or editions of cars are more popular in some countries than others. For example, most cars in the United States are automatic and diesel cars are more popular in some European countries than in the UK; assignments can be modified accordingly. It is possible to use other data from the *What Car* web site, such as luggage space or pulling power, and the assignments could easily be adapted to include these in the modelling process.

Time spent on data collection can be reduced by asking students to share part of the data they collect but still keeping their complete data different, although group assignments are also possible.

Many other tasks could be set, based on the available data; for example, regression analysis can be used to compare the retail prices of new cars with the target price given in *What Car*. There are also web sites that can be used to get car insurance quotes or car hire charges. Students could get quotations for some cars and carry out a regression analysis using these. The ideas introduced in this chapter would be relevant to such a scenario. A more ambitious project could explore most of the ideas in this chapter based on collecting larger data sets.

References

Aliaga, M. (2005) Teaching interactive statistics to understanding, in *Innovations in Teaching Statistics*, (ed. J. Garfield), Mathematical Association of America, Washington, D.C., pp. 75–81.

Altman, D. G. (1994) The scandal of poor medical research. *British Medical Journal*, **308**, 283–284.

American Statistical Association (ASA) (2007) *Guidelines for Assessment and Instruction in Statistics Education* (GAISE), http://www.amstat.org/education/gaise/ (accessed 11 December 2009).

American Statistical Association (2008) *Strategic Plan for Education*, http://www.amstat.org/about/index.cfm?fuseaction=strategicplan (accessed 11 December 2009).

Anderson, G. and Boud, D. (1996) Extending the role of peer learning in university courses. *Research and Development in Higher Education*, **19**, 15–19.

Anderson, L. W. and Krathwohl, D. R. (2001) A Taxonomy for Learning, Teaching, and Assessing: A Revision of Bloom's Taxonomy of Educational Objectives, Longmans, New York.

Angelo, T. (1999) Doing assessment as if learning matters most. *Bulletin of the American Association for Higher Education*, http://www.aacsb.edu/resource_centers/assessment/Angelo-TA-Reprint.asp (accessed 12 December 2009).

Arbuckle, J. L. (2008) *AMOS User's Guide* 17.0, SPSS Inc.

Barnett, V. (2004) Review of online statistics teaching material. *MSOR Connections*, **4**(2), 43–45, Learning and Teaching Support Network, Maths, Stats & OR Network, Birmingham, UK.

Barton, A., Van Duuren, M. and Haslam, P. (2006) Voluntary peer learning groups: Do students want more structure and are there any hard gains? *Psychology Learning and Teaching*, **5**(2), 146–152.

Barton, A., Van Duuren, M. and Haslam, P. (2007) Perceived social benefits of voluntary student collaboration. *Psychology Learning and Teaching*, **6**(1), 26–33.

Bass, K. M. and Glaser, R. (2004) Developing Assessments to Inform Teaching and Learning, *Center for the Study of Evaluation, Report No. 628*. National Center for Research on Evaluation, Standards and Student Testing, Los Angeles, CA.

Batt, J. (2004) Stolen Innocence: The Sally Clark Story – A Mother's Fight for Justice, Ebury Press, London.

Baumgartner, E. (2004) Student poster sessions. *The Science Teacher*, **71**(3), 1–4.

Assessment Methods in Statistical Education Edited by Penelope Bidgood, Neville Hunt and Flavia Jolliffe

Begg, A. (1997) Some emerging influences underpinning assessment in statistics, in *The Assessment Challenge in Statistics Education*, (eds I. Gal and J. B. Garfield), IOS Press, Amsterdam, Netherlands, pp. 17–25, http://www.stat.auckland.ac.nz/~iase/publications/assessbk/chapter02.pdf (accessed 11 December 2009).

Begg, A., Erickson, T., MacGillivray, H. and Matis, T. (2004) Working Group Report on Statistics Curriculum: Content and Framing, in *Curricular Developments in Statistical Education*, (eds G. Burrill and M. Camden), International Statistical Institute, Voorburg, The Netherlands, pp. 275–277.

Ben-Zvi, D. and Garfield, J. (2004a) Statistical literacy, reasoning and thinking: Goals, definitions, and challenges, in *The Challenge of Developing Statistical Literacy, Reasoning and Thinking*, (eds D. Ben-Zvi and J. Garfield), Kluwer Academic Publishers, Dordrecht, The Netherlands, pp. 3–16.

Ben-Zvi, D. and Garfield, J. (eds) (2004b) *The Challenge of Developing Statistical Literacy, Reasoning and Thinking*, Kluwer Academic Publishers, Dordrecht, The Netherlands.

Best, J. (2008) Birds – dead and deadly: Why numeracy needs to address social construction. *Numeracy*, **1**(1), article 6, http://services.bepress.com/numeracy/vol1/iss1/art6/(accessed 11 December 2009).

Betts, S. C. (2006) A test of prospect theory in the used car market: The non-linear effects of age and reliability on price. *Academy of Marketing Studies Journal*, **10**(2), 57–75.

Bidgood, P. and Cox, W. (2002) Student Assessment in MSOR. *MSOR Connections*, **2**(4), 9–13.

Bidgood, P., Hunt, D. N., Payne, B. and Simonite, V. (2007) *The Plagiarism in Statistics Assessment Project*, http://www.jiscpas.ac.uk/documents/pisa.pdf (accessed 11 December 2009).

Biehler, R. (2007) Assessing students' statistical competence by means of written reports and project work, in *Proceedings of the IASE Satellite Conference on Assessing Student Learning in Statistics*, (eds B. Phillips and K. L. Weldon), International Statistical Institute, Voorburg, The Netherlands, http://www.stat.auckland.ac.nz/~iase/publications/sat07/Biehler.pdf (accessed 11 December 2009).

Biggs, J. (1996) Enhancing teaching through constructive alignment. *Higher Education*, **32**(3), 347–364.

Biggs, J. (1999a) What the student does: Teaching for enhanced learning. *Higher Education Research and Development*, **18**(1), 57–75.

Biggs, J. (1999b) *Teaching for Quality Learning at University*, Open University Press, Buckingham.

Biggs, J. (2003) *Teaching for Quality Learning at University*, 2nd edn, Society for Research into Higher Education, Open University Press, Buckingham.

Bilgin, A. and Fraser, S. (2007) Empowering students to be the judges of their own performance through peer assessment, in *Proceedings of the IASE Satellite Conference on Assessing Student Learning In Statistics*, IASE, Guimarães, Portugal, 19–22 August, http://www.stat.auckland.ac.nz/~iase/publications/sat07/Bilgin_Fraser.pdf (accessed 11 December 2009).

Billing, D. (2007) Teaching for transfer for core/key skills in higher education: Cognitive skills. *Higher Education*, **53**, 483–516.

Bingham, R. (2001) *Assessment Criteria – A Guide*, Learning and Teaching Institute, Sheffield Hallam University.

Black, P. (2005) Assessment for learning. Where is it now and where is it going? Presented at Improving Student Learning through Assessment Conference, London, September, http://www.brookes.ac.uk/services/ocsld/isl/isl2005/ (accessed 11 December 2009).

Black, P. and Wiliam, D. (1998) Assessment and classroom learning. *Assessment in Education*, **5**(1), 7–74.

Blanco, M., Ginovart, M., Estela, M. R. and Jarauta, E. (2006) Teaching and learning mathematics and statistics at an Agricultural Engineering School, *Proceedings of the CIEAEM 58 Congress – Changes in Society: A challenge for mathematics education*, University of West Bohemia, Plzen, pp. 152–157.

Bloom, B. S. (1956) Taxonomy of Educational Objectives:The Classification of Educational Goals, Susan Fauer Company, Inc., Chicago.

Boud, D. (1995) Enhancing Learning through Self-assessment, Kogan Page, London.

Boud, D. (2000) Sustainable assessment: Rethinking assessment for the learning society. *Studies in Continuing Education*, **22**(2), 151–167.

Boud, D., Cohen, R. and Sampson, J. (1999) Peer learning and assessment. *Assessment and Evaluation in Higher Education*, **24**(4), 413–426.

Boud, D., Cohen, R. and Sampson, J. (eds) (2001) Peer Learning in Higher Education: Learning from and with each Other, Kogan Page, London.

Boynton, P. (2004) Teaching statistics – the missing ingredients. *Radical Statistics*, **87**(e), 19–30, http://www.radstats.org.uk/no087/index.htm (accessed 11 December 2009).

Brew, A. (2006) *Research and Teaching: Beyond the Divide*, Palgrave Macmillan, Basingstoke.

Brew, A. and Boud, D. (1995) Teaching and research: Establishing the vital link in learning. *Higher Education*, **29**(3), 261–273.

Broers, N. (2006) *Learning Goals: The Primacy of Statistical Knowledge*, ICOTS-7. http://www.stat.auckland.ac.nz/~iase/publications/17/6G2_BROE.pdf (accessed 11 December 2009).

Brown, G. (2001) *Assessment: A Guide for Lecturers*, LTSN Generic Centre Assessment Series No 3, York.

Brown, J. (2007) Teaching sampling statistics: Experience from the MSc in Official Statistics Programme, International Statistical Institute, 56th Session. International Statistical Institute, http://www.stat.auckland.ac.nz/~iase/publications/isi56/IPM44_Brown.pdf (accessed 11 December 2009).

Brown, S. and Knight, P. (1994) *Assessing Learners in Higher Education*, Kogan Page, London.

Budgett, S. and Pfannkuch M. (2007) *Assessing Students' Statistical Literacy*. International Association for Statistical Education (IASE/ISI Satellite, 2007 Portugal), http://www.StatLit.org/pdf/2007BudgettPfannkuchIASE.pdf (accessed 11 December 2009).

Bulmer, M. and Low, E. (2008) Technology for insight into students' beliefs about statistics in large classes, in *Proceedings 6th Australian Conference on Teaching*

Statistics, (eds H. MacGillivray and M. Martin), pp. 104–109, http://sky.scitech.qut.edu.au/~macgilli/ozcots2008/OZCOTS-08-Proceedings.doc (accessed 11 December 2009).

Cengage Learning (2008) *CengageNow*, http://cvg.ilrn.com/ilrn/ (accessed 11 December 2009).

Centre for Inquiry-based Learning in the Arts and Social Sciences (2006) *PEBBLE Psychological Enquiry-Based Learning*, University of Sheffield, Sheffield, http://www.shef.ac.uk/cilass/projects/psychol.html (accessed 11 December 2009).

Chance, B. L. (1997) Experiences with authentic assessment techniques in an introductory statistics course. *Journal of Statistics Education*, **5**(3), http://www.amstat.org/publications/jse/v5n3/chance.html (accessed 11 December 2009).

Chance, B. L. (2002) Components of statistical thinking and implications for instruction and assessment. *Journal of Statistics Education*, **10**(3), http://www.amstat.org/publications/jse/v10n3/chance.html (accessed 26 December 2008).

Chance, B. L. (2005) Integrating pedagogies to teach statistics, in *Innovations in Teaching Statistics*, (ed. J. Garfield), Mathematical Association of America, Washington, D.C., pp. 101–110.

Cobb, G. (1992) Teaching statistics, in *Heeding the Call for Change: Suggestions for Curricular Action*, (ed. L. Steen), Mathematical Association of America Notes and Reports Series 22, 3–33.

Cobb, G. (2007) One possible frame for thinking about experiential learning. *International Statistical Review*, **75**(3), 336–347.

Conners, F. A., McCowan, S. M. and Roskos-Ewoldseon, B. (1998) Unique challenges in teaching undergraduates statistics. *Teaching of Psychology*, **25**(1), 41–42.

Cruise Scientific (2008) *Visual Statistics Studio*, http://www.visualstatistics.net (accessed 11 December 2009).

Conway, R., Kember, D., Sivan, A. and Wu, M. (1993) Peer assessment of an individual's contribution to a group project. *Assessment and Evaluation in Higher Education*, **18**(1), 45–56.

Cox, B. R. (1997) The rediscovery of the active learner in adaptive contexts: A developmental-historical analysis of transfer. *Educational Psychologist*, **32**(1), 41–55.

Dahlgren, M. Albrandt and Dahlgren, L. (2002). Portraits of PBL: Students' experiences of the characteristics of problem-based learning in physiotherapy, computer engineering and psychology. *Instructional Science*, **30**(2), 111–127.

Darius, P., Portier, K. and Schrevens, E. (2007) Virtual experiments and their use in teaching experimental design. *International Statistical Review*, **75**(3), 281–294.

Davies, N. and Payne, B. (2001) Web-created real data worksheets. *MSOR Connections*, **1**(4), 15–17, http://mathstore.gla.ac.uk/headocs/14webcreatedrealdata.pdf (accessed 11 December 2009).

delMas, R. (2002) Statistical literacy, reasoning, and thinking: A commentary. *Journal of Statistics Education*, **10**(3), http://www.amstat.org/publications/jse/v10n3/delmas_discussion.html (accessed 11 December 2008).

delMas, R., Garfield, J., Chance, B. L. (1999) A model of classroom research in action: Developing simulation activities to improve students' statistical reasoning. *Journal of Statistics Education*, **7**(3), http://www.amstat.org/publications/jse/secure/v7n3/delmas.cfm (accessed 11 December 2009).

delMas, R., Garfield, J., Ooms, A., Chance, B. L. (2006) Assessing students' conceptual understanding after a first course of statistics. *Annual Meetings of the American Educational Research Association*, San Francisco, CA, https://app.gen.umn.edu/artist/articles/AERA_2006CAOS.pdf (accessed 11 December 2009).

delMas, R., Garfield, J., Ooms, A. and Chance, B. (2007) Assessing students' conceptual understanding after a first course in statistics. *Statistics Education Research Journal*, **6**(2), 28–58, http://www.stat.auckland.ac.nz/~iase/serj/SERJ6(2)_delMas.pdf (accessed 11 December 2008).

Delucchi, M. (2007) Assessing the impact on group projects on examination performance in social statistics. *Teaching in Higher Education*, **2**(4), 447–460.

Downing, S. M. (2006) Twelve steps for effective test development, in *Handbook of test development*, (eds S. M. Downing and T. M. Haladyna), Lawrence Erlbaum Associates, Inc., Mahwah, NJ, pp. 3–26.

Economic and Social Research Council (ESRC). (2006) Invitation for expressions of interest: International bench-marking review of best practice in the provision of undergraduate teaching in quantitative methods in the social sciences.

Engers, M., Hartmann, M. Stern, S. (2009) Mileage drives used car prices. *Journal of Applied Econometrics*, **24**(1), 1–33.

Evans, R. (2005) Active learning and technology in the community college classroom, in *Innovations in Teaching Statistics*, (ed. J. Garfield), Mathematical Association of America, Washington, D.C., pp. 65–71.

Felder, R. and Spurlin, J. (2005) Applications, reliability and validity of the index of learning styles. *International Journal of Engineering Education*, **21**(1), 103–112.

Field, A. P. (2009a) *Discovering Statistics Using SPSS*, 3rd edn, Sage, London.

Field, A. P. (2009b) Can humour make students love statistics? *The Psychologist*, **22**(3), 210–213.

Fisher, M. and Moore, S. (2005) Enquiry-based learning links psychology theory to practice. *British Journal of Midwifery*, **13**(3), 148–152.

Forster, M. and Smith, D. P. (2007) Assessing large second year undergraduate service courses in data analysis, in *Proceedings of the IASE Satellite Conference on Assessing Student Learning in Statistics*, (eds B. Phillips and K. L. Weldon), International Statistical Institute, Voorburg, The Netherlands, http://www.stat.auckland.ac.nz/~iase/publications.php?show=sat07 (accessed 11 December 2009).

Forster, M., Smith, D. P. and Wild, C. J. (2005) Teaching students to write about statistics, in *Proceedings of the IASE Satellite Conference on Statistics Education and the Communication of Statistics*, (eds B. Phillips and K. L. Weldon), International Statistical Institute, Voorburg, The Netherlands, http://www.stat.auckland.ac.nz/~iase/publications/14/forster.pdf (accessed 11 December 2009).

Francis, G. (2005) An approach to report writing in statistics courses, in *Proceedings of the IASE Satellite Conference on Statistics Education and the Communication of Statistics*, (eds B. Phillips and K. L. Weldon), International Statistical Institute, Voorburg, The Netherlands, http://www.stat.auckland.ac.nz/~iase/publications/14/francis.pdf (accessed 11 December 2009).

Frankcom, G. (2007) *Statistics teaching and learning: The New Zealand experience*, http://tsg.icme11.org/document/get/489 (accessed 11 December 2008).

GAISE group (2005) *Guidelines for Assessment and Instruction in Statistics Education (GAISE) College Report*, http://www.amstat.org/education/gaise/GAISECollege.htm (accessed 11 December 2009).

Gal, I. (2002) Adult's statistical literacy: Meanings, components, responsibilities. *International Statistical Review* **70**(1), 1–51, International Statistical Institute, http://www.stat.auckland.ac.nz/~iase/cblumberg/gal.pdf (accessed 11 December 2009).

Gal, I. (2003) Teaching for statistical literacy and services of statistical agencies. *The American Statistician*, **57**(2), 80–84, http://pubs.amstat.org/toc/tas/57/2 (accessed 11 December 2009).

Gal, I. and Garfield, J. (1997a) Curricular goals and assessment challenges in statistics education, in *The Assessment Challenge in Statistics Education*, (eds Gal, I. and J. B. Garfield) IOS Press, Amsterdam, Netherlands, 1–14, http://www.stat.auckland.ac.nz/~iase/publications/assessbk/chapter01.pdf (accessed 11 December 2009).

Gal, I. and Garfield, J. (1997b) *The Assessment Challenge in Statistics Education*, IOS Press, Amsterdam, Netherlands, http://www.stat.auckland.ac.nz/~iase/publications/assessbk/(accessed 11 December 2009).

Gal, I., Ginsberg, L. and Schau, C. (1997) Monitoring attitudes and beliefs in statistics education, in *The Assessment Challenge in Statistics Education*, (eds I. Gal and J. Garfield), 37–54, IOS Press, Amsterdam, Netherlands, http://www.stat.auckland.ac.nz/~iase/publications/assessbk/chapter04.pdf (accessed 11 December 2009).

Gambino, J. and Gough, H., (2005) Teaching sampling in a government statistical agency: The Canadian experience. *International Statistical Institute, 55th Session 2005*, International Statistical Institute, http://www.stat.auckland.ac.nz/~iase/publications/13/Gambino-Gough.pdf (accessed 11 December 2009).

Garfield, J. B. (1994) Beyond testing and grading: Using assessment to improve student learning. *Journal of Statistics Education*, **2**(1).

Garfield, J. (1995) How students learn statistics. *International Statistical Review*, **63**(1), 25–34.

Garfield, J. (2002) The challenge of developing statistical reasoning. *Journal of Statistics Education*, **10**(3), http://www.amstat.org/publications/jse/v10n3/garfield.html (accessed 11 December 2009).

Garfield, J. (2003) Assessing statistical reasoning. *Statistics Education Research Journal*, **2**(1), 22–38, http://www.stat.auckland.ac.nz/%7Eiase/serj/SERJ2(1).pdf (accessed 11 December 2009).

Garfield, J. (ed.) (2005) *Innovations in Teaching Statistics*, Mathematical Association of America, Washington, D.C., pp. 39–47.

Garfield, J. and Ben-Zvi, D. (2008) Developing Students' Statistical Reasoning: Connecting Research and Teaching Practice, Springer Publishers, The Netherlands.

Garfield, J. and delMas, R. (2010) A website that provides resources for assessing students' statistical literacy, reasoning, and thinking. *Teaching Statistics*, **32**(1), 2–7.

Garfield, J., delMas, R. and Chance, B. (2003) The web-based ARTIST: Assessment Resource Tools for Improving Statistical Thinking. American Educational Research Association meeting, Chicago, April 2003, https://app.gen.umn.edu/artist/articles/AERA_2003.pdf (accessed 11 December 2009).

Garfield, J. and Gal, I. (1999) Assessment and statistics education: Current challenges and directions. *International Statistical Review* **67**, 1–12.

Garfield, J., Hogg, R. V., Schau, C. and Whittinghill, D. (2002) First courses in statistical science: The status of educational reform efforts. *Journal of Statistics Education* **10**(2), http://www.amstat.org/publications/jse/v10n2/garfield.html (accessed 11 December 2009).

Gatfield, T. (1999) Examining student satisfaction with group projects and peer assessment. *Assessment and Evaluation in Higher Education*, **24**(4), 365–77.

Gelman, A. and Nolan, D. (2002) *Teaching statistics: A bag of tricks*, Oxford University Press, New York.

Gerstman, B. B. (2008) *StatPrimer*, http://www.sjsu.edu/faculty/gerstman/StatPrimer (accessed 11 December 2009).

Gibbs, G. and Simpson, C. (2004) Does your assessment support your students' learning? *Journal of Learning and Teaching in Higher Education*, **1**(1), 3–31.

Gibson, L., Marriott, J. and Davies, N. (2007) Solving the problem of teaching statistics? *International Statistical Institute, 56th Session, 2007*. International Statistical Institute. www.stat.auckland.ac.nz/~iase/publications/isi56/CPM80_Gibson.pdf (accessed 12 December 2009).

Gigerenzer, G., Todd, P. and ABC Research Group (eds) (1999) *Simple Heuristics that Make us Smart*, Oxford University Press, New York.

Gilman, R. (2006) *Current Practices in Quantitative Literacy*, Mathematical Association of America (MAA).

Gnaldi, M. (2006) The relationship between poor numerical abilities and subsequent difficulty in accumulating statistical knowledge. *Teaching Statistics*, **28**, 49–53.

Goldacre, B. (2006) When the facts get in the way of a story. *The Guardian*, http://www.guardian.co.uk/science/2006/apr/01/badscience.uknews (accessed 11 December 2009).

Goldacre, B. (2008) *Bad Science*, 4th Estate, London.

Goldfinch, J. (1994) Further developments in peer assessment of group projects. *Assessment & Evaluation in Higher Education*, **19**(1), 29–35.

Goldring, L. and Wood, J. (2007) *A Guide to the Facilitation of Enquiry-Based Learning for Graduate Students*, University of Manchester, Centre for Excellence in Enquiry-Based Learning, Manchester, http://www.campus.manchester.ac.uk/ceebl/resources/evaluation/guide_to_fac_v1_bookletlayout.pdf (accessed 11 December 2009).

Gollwitzer, P. M. (1999) Implementation intentions: Strong effects of simple plans. *American Psychologist*, **54**, 493–503.

Gordon, I., Finch, S. and Maillardet, R. (eds) (2008) Statistics as breadth: The Melbourne experiment, in *Proceedings of the 6th Australian Conference on Teaching Statistics*, (H. MacGillivray and M. Martin), pp. 143–148, http://sky.scitech.qut.edu.au/~macgilli/ozcots2008/OZCOTS-08-Proceedings.doc (accessed 11 December 2009).

Gould, R., Kreuter, F. and Palmer, C. (2006) Towards statistical thinking: Making data real, in *Proceedings of the International Conference on Teaching Statistics 7*, http://www.stat.auckland.ac.nz/~iase/publications/17/7A2_GOUL.pdf (accessed 11 December 2009).

Harvey, A. J. (2010) The merits of a general numeracy test as a predictor of undergraduate statistics performance. *Psychology Learning and Teaching*, **8**(2), 16–22.

Haubl, G. (1996) A cross-national investigation of the effects of country of origin and brand name on the evaluation of a new car. *International Marketing Review*, **13**(5), 70–97.

Hayden, R. (1989) Using writing to improve student learning of statistics. *Writing Across the Curriculum*, **1**(1), 3–9.

Hayden, R. (2004) Planning a statistical literacy program at the college level: Musings and a bibliography. *ASA Proceedings of the Section on Statistical Education*, [CD-ROM] 2685–2692, http://www.statlit.org/PDF/2004HaydenASA.pdf (accessed 11 December 2009).

Hilton, S. C., Schau, C. and Olsen, J. A. (2004) Survey of attitudes toward statistics: Factor structure invariance by gender and by administration time. *Structural Equation Modeling*, **11**, 92–109.

Holmes, P. (1997) Assessing project work by external examiners, in *The Assessment Challenge in Statistics Education*, (eds I. Gal and J. B. Garfield), IOS Press, Amsterdam, http://www.stat.auckland.ac.nz/~iase/publications/assessbk/chapter12.pdf (accessed 11 December 2009).

Holmes, P. (2002) Teaching, learning and assessment: Complementary or conflicting categories for school statistics. Sixth International Conference on Teaching Statistics (ICOTS-6), Cape Town, South Africa, 7–12 July 2002, http://www.stat.auckland.ac.nz/~iase/publications/1/04_ho.pdf (accessed 16 December 2009).

Hoyles, C., Wolf, A., Molyneaux-Hodgson, S. and Kent, P. (2002) *Mathematical Skills in the Workplace*. Final report to the Science, Technology and Mathematics (STM) Council. Institute of Education, University of London, and STM Council, London.

Hubbard, R. (2003) Questions for formative and summative assessment that encourage deep rather than surface approaches to learning basic statistics in a computer environment. Paper presented at the joint Statistical Meetings, https://app.gen.umn.edu/artist/articles/hubbard.pdf (accessed 11 December 2008).

Huey, D. (2001) The potential utility of problem-based learning in the education of clinical psychologists and others. *Education for Health*, **14**(1), 11–19.

Hulme, P., Sherratt, T. and Moss, J. (2008) *StatWeb*, http://www.dur.ac.uk/stat.web/ (accessed 11 December 2009).

Hunt, D. N. (2005) Using Microsoft Office to generate individualized tasks for students. *Teaching Statistics*, **27**(2), 45–48.

Hunt, D. N. (2007a) Individualized statistics coursework using spreadsheets. *Teaching Statistics*, **29**(2), 38–43.

Hunt, D. N. (2007b) Setting personalized homework exercises using ISCUS. *Teaching Statistics*, **29**(3), 66–70.

Hutchings, W. (2006) Facilitating enquiry-based learning: Some digressions. Keynote lecture. 2nd Southern Universities EBL Network Event, 11 January 2006, University of Surrey. University of Manchester, Centre for Excellence in Enquiry Based Learning, Manchester, http://www.campus.manchester.ac.uk/ceebl/resources/papers/surreyjan06_keynote.pdf (accessed 11 December 2009).

Huws, N., Reddy, P. and Talcott, J. (2006) Predicting university success in psychology: Are subject-specific skills important? *Psychology Learning and Teaching*, **5**(2), 133–140.

International Institute for Population Sciences (2009) National Family Health Survey, India, http://www.nfhsindia.org/ (accessed 11 December 2009).

Isaacson, M. (2005) *Statistical Literacy – an Online Course at Capella University*, http://www.statlit.org/pdf/2005IsaacsonASA.pdf (accessed 11 December 2009).

James, R., McInnis, C. and Devlin, M. (2002) *Assessing Learning in Australian Universities*, The University of Melbourne Centre for the Study of Higher Education, Melbourne.

Jara-Peña, E., Villegas, A. and Sánchez P. (2002) Contenido de N, P, K y rendimiento de frambuesa roja (Rubus idaeus L.) 'Autumn Bliss' orgánico asociada con lupino (Lupinus mutabilis Sweet). *Revista Peruana de. Biologia*, **9**(2), 84–93.

Jenkins, A. and Healey, M. (2005) *Institutional Strategies to Link Teaching and Research*. The Higher Education Academy, York, http://www.heacademy.ac.uk/resources/detail/id585_institutional_strategies_to_link_teaching_and_research (accessed 11 December 2009).

Jenkins, A., Breen, R., Lindsay, R. and Brew, A. (2003) *Reshaping Teaching in Higher Education: Linking Teaching and Research*, Routledge, London.

Johnson, D. W., Johnson, R. T. and Houlbec, E. (1998) *Cooperation in the Classroom* (6th edn), Interaction Book company, Edina, MN.

Johnson, D. W. and Johnson, R. T. (2002) *Meaningful Assessment: A Manageable and Cooperative Process*. Allyn and Bacon, Boston, MA.

Johnson, D. W. and Johnson, F. P. (2006) *Joining Together: Group Theory and Group Skills* (9th edn), Allyn and Bacon, Boston, MA.

Jolliffe, F. (1997) Issues in constructing assessment instruments for the classroom, in *The Assessment Challenge in Statistics Education*, (eds I. Gal and J. B. Garfield), IOS press, Amsterdam, pp. 191–204, http://www.stat.auckland.ac.nz/~iase/publications/assessbk/chapter15.pdf (accessed 11 Dember 2009).

Jolliffe, F. (2003) The same but different – how to introduce variation within computing assessment tasks. *Teaching Statistics*, **25**(1), 14–16.

Jolliffe, F. (2007) The changing brave new world of statistics assessment. IASE/ISI Satellite conference on Assessing Student Learning in Statistics, Guimarães, Portugal, August 2007. http://www.stat.auckland.ac.nz/~iase/publications/sat07/Jolliffe.pdf (accessed 14 September 2009).

Juwah, C., Macfarlane-Dick, D., Matthew, B. *et al.* (2004) *Enhancing Student Learning through Effective Formative Feedback*. The Higher Education Academy, http://www.heacademy.ac.uk/assets/York/documents/resources/resourcedatabase/id353_senlef_guide.pdf (accesssed 18 September 2006).

Kahn, P. and O'Rourke, K. (2005) Understanding enquiry-based learning, in *Handbook of Enquiry and Problem Based Learning*, (eds T. Barrett, I. Mac Labhrainn and H. Fallon), CELT, Galway, http://www.aishe.org/readings/2005-2/chapter1.pdf (accessed 11 December 2009).

Kanji, G. K. (1979) The role of projects in statistical education. *The Statistician*, **28**, 19–27.

Kelly, A. E., Sloane, F. and Whittaker, A. (1997) Simple approaches to assessing underlying understanding of statistical concepts, in *The Assessment Challenge in Statistics Education*, (eds I. Gal and J. B. Garfield), IOS Press, Amsterdam, pp. 85–90, http://www.stat.auckland.ac.nz/~iase/publications/assessbk/chapter07.pdf (accessed 11 December 2009).

Kooreman, P. and Hann, M. A. (2006) Price anomalies in the used car market. *De Economist*, **154**(1), 41–62.

Kulpers, S., Cannegieter, S. C., Middeldorp, S. *et al.* (2007) The absolute risk of venous thrombosis after air travel: A cohort study of 8,755 employees of international organisations. *PLoS Medicine*. **4**, 9.

Kvam, P. H. (2000) The effect of active learning methods on student retention in Engineering Statistics. *The American Statistician* **54**(2), 136–140.

Lane, D. (2008) *Hyperstat*, http://davidmlane.com/hyperstat/questions/index.html (accessed 11 December 2009).

Lee, C. (2005) Using the PACE strategy to teach statistics, in *Innovations in Teaching Statistics*, (ed. J. Garfield), Mathematical Association of America, Washington, D.C., pp. 13–22.

Lill, M. and Wilkinson, T. J. (2005) Judging a book by its cover: A descriptive survey of patients' preference for doctors' appearance and mode of dress. *British Medical Journal*, **331**, 1524–1527.

Lipson, K. and Kokonis, S. (2005) The implications of introducing report writing into an introductory statistics subject, in *Proceedings of the IASE Satellite Conference on Statistics Education and the Communication of Statistics*, (eds B. Phillips and K. L. Weldon), International Statistical Institute, Voorburg, The Netherlands. http://www.stat.auckland.ac.nz/~iase/publications/14/lipson.pdf (accessed 11 December 2009).

Liu, N-F. and Carless, D. (2006) Peer Feedback: The learning element of peer assessment. *Teaching in Higher Education*, **11**(3), 279–290.

Livingstone, D. and Lynch, K. (2000) Group project work and student-centred active learning: Two different experiences. *Studies in Higher Education* **25**(3), 325–345.

Lock, R. H. (1993) 1993 New car data. *Journal of Statistics Education*, **1**, 1, http://www.amstat.org/publications/jse/v1n1/datasets.lock.html (accessed 11 December 2009).

Lutsky, N. (2006) Quirks of rhetoric: A quantitative analysis of quantitative reasoning in student writing. *ASA Proceedings of the Section on Statistical Education*, pp. 2319–2322. http://www.StatLit.org/pdf/2006LutskyASA.pdf (accessed 11 December 2009).

MacGillivray, H. L. (1998) Developing and synthesizing statistical skills for real situations through student projects, in *Proceedings of the 5th International Conference on Teaching Statistics*, pp. 1149–1155, International Statistical Institute, Voorburg, Netherlands. http://www.stat.auckland.ac.nz/~iase/publications/2/Topic8n.pdf.

MacGillivray, H. L. (2002) One thousand projects. *MSOR Connections*, **2**(1), 9–13.

MacGillivray, H. (2004) Coherent and purposeful development in statistics across the educational spectrum, in *Curricular Developments in Statistical Education*, (eds G. Burrill and M. Camden), International Statistical Institute, Voorburg, The Netherlands.

MacGillivray, H. (2005) Helping students find their statistical voices, in *Proceedings of the IASE Satellite Conference on Statistics Education and the Communication of Statistics*, (eds B. Phillips and K. L. Weldon), International Statistical Institute, Voorburg, The Netherlands, http://www.stat.auckland.ac.nz/~iase/publications/14/macgillivray.pdf (accessed 11 December 2009).

MacGillivray, H. L. (2007) Weaving assessment for student learning in probabilistic reasoning at the introductory tertiary level, in *Proceedings IASE/ISI Conference on Assessing Student Learning in Statistics, Portugal*, (eds B. Chance and B. Phillips), International Statistical Institute, Voorburg, The Netherlands, http://www.stat.auckland.ac.nz/~iase/publications/sat07/Macgillivray.pdf (accessed 11 December 2009).

MacGillivray, H. L. (2008) Examples from an introductory course in developing probabilistic statistical thinking: Part 1. *MSOR Connections* **8**(1), 7–10.

MacKay, R. J. and Oldfield, W. (1994) *Stat 231 Course Notes, Fall 1994*, University of Waterloo.

Mackisack, M. (1994) What is the use of experiments conducted by statistics students? *Journal of Statistics Education* **2**, 1. http://www.amstat.org/publications/jse/v2n1/mackisack.html (accessed 11 December 2009).

Macnaughton, D. (2004) The introductory statistics course: The entity-property-relationship approach, http://www.MatStat.com/teach (accessed 11 December 2009).

Madison, B. L. (2006) Pedagogical challenges of quantitative literacy. *2006 Proceedings of the ASA Section on Statistical Education*, [CD-ROM] pp. 2323–2328, http://www.StatLit.org/pdf/2006MadisonASA.pdf (accessed 11 December 2009).

Madison, B. L. and Steen L. A. (2008) *Calculation vs. Context: Quantitative Literacy and its Implications for Teacher Education*, Mathematical Association of America (MAA), http://www.maa.org/ql/calcvscontext.html (accessed 11 December 2009).

Marriott, J., Davies, N. and Gibson, L. (2009) Teaching, learning and assessing statistical problem solving. *Journal of Statistics Education* **17**, 1, http://www.amstat.org/publications/jse/v17n1/marriott.html (accessed 11 December 2009).

Martin, M. A. (2003) "It's like... you know": The use of analogies and heuristics in teaching introductory statistical methods. *Journal of Statistics Education*, **11**(2). http://www.amstat.org/publications/jse/v11n2/martin.html.

Mathematics, Statistics and Operational Research Overview Report (2000) *Quality Assurance Agency for Higher Education*, Quality Assurance Agency, Gloucester, UK.

Matt, E. G. (2008) Tobacco use and asking prices of used cars, *Tobacco Induced Diseases*, http://www.pubmedcentral.nih.gov/articlerender.fcgi?artid=2547891 (accessed 11 December 2009).

McColl, J., Payne, B. and MacGillivray, H. L. (2007) Model choice on the web. Presented at CETL-MSOR Conference, Birmingham, UK, September, 2007.

McIntosh, A., Reys, B., and Reys, R. (1992) A proposed framework for examining number sense. *For the Learning of Mathematics*, **12**(3), 25–31.

McKenzie, J. D., Jr. (2004) Conveying the core concepts. *ASA 2004 Proceedings of the Section on Statistical Education*, [CD-ROM] 2755–2757, http://www.statlit.org/pdf/2004McKenzieASA.pdf (accessed 11 December 2009).

McLaren, H. and B. J. McLaren (2003) Electric bill data. *Journal of Statistics Education*, **11**, 1, http://www.amstat.org/publications/jse/v11n1/datasets.mclaren.html (accessed 11 December 2009).

McLeod, I., Zhang, Y., and Yu, H. (2003) Multiple-choice randomization. *Journal of Statistics Education*, **11**, 1, http://www.amstat.org/publications/jse/v11n1/mcleod.html (accessed 11 December 2009).

Mead, R. (1990) Statistical Games 1 – Tomato. *Teaching Statistics*, **12**, 76–78.

Meldrum, M. (1998) A calculated risk: the Salk polio vaccine field trials of 1954. *British Medical Journal*, **317**, 1233–1236, http:www.bmj.com/cgi/content/full/317/7167/1233 (accessed 11 December 2009).

Merriman, L. (2006) Using media reports to develop statistical literacy in Year 10 students, in *Proceedings of the 7th International Conference on Teaching Statistics*. (eds A. Rossman and B. Chance), [CD-ROM], International Statistical Institute, Voorburg, The Netherlands, http://www.stat.auckland.ac.nz/~iasc/publications/17/8A3_MERR.pdf (accessed 11 December 2009).

Meyers, C. and Jones, T. B. (1993) *Promoting Active Learning: Strategies for the College Classroom*, Jossey-Bass Publishers, San Francisco.

Miller, G., Tybur, J. M. and Jordan, B. D. (2007) Ovulatory cycle effects on tip earnings by lap dancers: Economic evidence for human estrus? *Evolution and Human Behavior*, **28**, 375–381.

Miller, J. E. (2004) *The Chicago Guide to Writing about Numbers*, The University of Chicago Press, Chicago, IL.

Miller, J. E. (2005) *The Chicago Guide to Writing about Multivariate Analysis*, The University of Chicago Press, Chicago, IL.

Mislevy, R. J., Steinberg, L. S., and Almond, R. G. (2003) On the structure of educational assessments. *Measurement: Interdisciplinary Research and Perspectives*, **1**(1), 3–62.

Moore, D. S. (1997) New pedagogy and new content: The case of statistics (with Discussion). *International Statistical Review*, **65**(2), 123–165, http://www.stat.auckland.ac.nz/~iase/publications/isr/97.Moore.pdf (accessed 11 December 2009).

Moore, D. S. (1998) Statistical literacy and statistical competence in the 21st Century. Abstract for his talk at a 'Making Statistics More Effective in Schools of Business' (MSMESB) conference in Iowa City, IA, http://www.StatLit.org/pdf/1998MooreMSMESB.pdf (accessed 11 December 2009).

Moreno, J. (2002) Toward a statistically literate citizenry: What statistics should everyone know. International Conference on Teaching Statistics (ICOTS-6). Singapore. http://www.stat.auckland.ac.nz/~iase/publications/1/1b6_more.pdf (accessed 11 December 2009).

Mulhern, G. and Wylie, J. (2004) Changing levels of numeracy and other core mathematical skills among psychology undergraduates between 1992 and 2002. *British Journal of Psychology*, **95**, 355–370.

Mulhern, G. and Wylie, J. (2006) Mathematical prerequisites for learning statistics in psychology: Assessing core skills of numeracy and mathematical reasoning among undergraduates. *Psychology Learning and Teaching*, **5**(2), 119–133.

Murch, S., Anthony, A., Casson, D. *et al.* (2004) Retraction of an interpretation. *The Lancet*, **363**(9411), 750.

Murdoch, J. and Barnes, J. A. (1974) *Basic Statistics Laboratory Instruction Manual*, Macmillan, London.

Murray, S. and Gal, I. (2002) Preparing for diversity in statistics literacy: Institutional and educational implications, in *Proceedings of the Sixth International Conference on Teaching Statistics*, (ed. B. Phillips) [CD-ROM], International Statistical Institute, Voorburg, The Netherlands, http://www.stat.auckland.ac.nz/~iase/publications/1/02_mu.pdf (accessed 12 December 2009).

National Research Council (2001) *Knowing what Students Know: The Science and Design of Educational Assessment*. Committee on the Foundations of Assessment. (eds J. Pelligriono, N. Chudowsky, and R. Glaser), National Academy Press, Washington, D.C.

Nicol, D. and Macfarlane-Dick, D. (2004) Rethinking formative assessment in HE: A theoretical model and seven principles of good feedback practice. The Higher Education Academy SENLEF Project, http://www.heacademy.ac.uk/ourwork/learning/assessment/senlef/principles (accessed 12 December 2009).

Niss, M. (1993) Assessment in mathematics education and its effects: An introduction, in Investigations into assessment in mathematics education, (ed. M. Niss), Kluwer Academic Publishers, Dordrecht, pp. 1–30.

Nolan, D. and Lang, D. T. (2007) Dynamic, interactive documents for teaching statistical practice. *International Statistical Review*, **75**(3), 295–321.

Obremski, T. (2008) Pricing models using real data. *Teaching Statistics*, **30**(2), 44–48.

O'Donovan B., Price, M., and Rust, C. (2004) Know what I mean? Enhancing student understanding of assessment standards and criteria. *Teaching in Higher Education*, **9**, 325–335.

O'Fallon, M. (2000) Undergraduate education and communication, *Amstat News*, **277**, 2–3.

O'Muircheartaigh, C. (2005) Balancing statistical theory, sampling concepts, and practicality in the teaching of survey sampling, in *Proceedings of the International Statistical Institute, 55th Session*, 2005, International Statistical Institute, http://www.stat.auckland.ac.nz/~iase/publications/13/O_Muircheartaigh.pdf (accessed 12 December 2009).

Onwuegbuzie, A. J. and Wilson, V. (2003) Statistics anxiety: Nature, etiology antecedents, effects, and treatments – a comprehensive review of the literature. *Teaching in Higher Education*, **8**, 195–209.

Pardoe, I. (2008) Modelling home prices using realtor data. *Journal of Statistics Education*, **16**, 2, http://www.amstat.org/publications/jse/v16n2/datasets.pardoe.html (accessed 12 December 2009).

Parke, C. S. (2008) Reasoning and communicating in the language of statistics. *Journal of Statistics Education*, **16**, 1, http://www.amstat.org/publications/jse/v16n1/parke.html (accessed 12 December 2009).

Pauli, R., Mohiyeddini, C., Bray, D., Michie, F. and Street, B. (2008) Individual differences in negative group work experiences in collaborative student learning. *Educational Psychology*, **28**(1), 47–58.

Peck, R. (2005) There's more to statistics than computation – teaching students how to communicate statistical results, in *Proceedings of the IASE Satellite Conference on Statistics Education and the Communication of Statistics*, (eds B. Phillips and K. L. Weldon), Voorburg, International Statistical Institute, The Netherlands, http://www.stat.auckland.ac.nz/~iase/publications/14/peck.pdf (accessed 12 December 2009).

Peck, R., Olsen, C. and Devore, J. (2008) *Introduction to Statistics and Data Analysis*, (3rd edn), Thompson, Belmont, California.

Perdisco (2008) http://www.perdisco.com.au/

Perkins, D. M and Salomon, G. (1989) Are cognitive skills context-bound? Educational Researcher **18**(10), 16–25.

Petersen, W. (2005) Teaching a CHANCE course, in *Innovations in Teaching Statistics*, (ed. J. Garfield), Mathematical Association of America, Washington, D.C., pp. 39–47.

Petocz, P. and Reid, A. (2007) Learning and assessment in statistics, IASE/ISI Satellite, 2007, International Statistical Institute, http://www.stat.auckland.ac.nz/~iase/publications/sat07/Petocz_Reid.pdf (accessed 12 Decmber 2009).

Pfannkuch, M. and Wild, C. (2004) Towards an understanding of statistical thinking, in *The Challenge of Developing Statistical Literacy, Reasoning and Thinking*, (eds D. Ben-Zvi and J. Garfield), Kluwer Academic Publishers, Dordrecht, The Netherlands, pp. 17–46.

Phillips, B. and Weldon, K. L. (2005) *Proceedings of the IASE Satellite Conference on Statistics Education and the Communication of Statistics*, International Statistical Institute, Voorburg, The Netherlands, http://www.stat.auckland.ac.nz/~iase/publications.php?show=14, (accessed 12 December 2009).

PiSA – Plagiarism in Statistics Assessment (2007), http://www.coventry.ac.uk/ec/~nhunt/pisa.pdf (accessed 12 December 2009).

Pond III, S. B. (2004) All in the Balance: Psychology 201 'Controversial Issues in Psychology', in *Teaching and Learning Through Inquiry. A Guidebook for Institutions and Instructors*, (ed. V. S. Lee), Stylus, Sterling, Virginia, pp. 31–40.

Prosser, M. and Trigwell, K. (1999) *Understanding Learning and Teaching: The Experience in Higher Education*, Society for Research into Higher Education and Open University Press, Buckingham.

Prvan, T. and Ascione, J. (2005) Enabling students to communicate statistical findings, in *Proceedings of the IASE Satellite Conference on Statistics Education and the Communication of Statistics*, (eds B. Phillips and K. L. Weldon), International Statistical Institute, Voorburg, The Netherlands, http://www.stat.auckland.ac.nz/~iase/publications/14/prvan.pdf. (accessed 12 December 2009).

Quality Assurance Agency (2007) *Benchmark Statement for Psychology*, http://www.qaa.ac.uk/academicinfrastructure/benchmark/statements/psychology07.asp (accessed 12 December 2009).

R Development Core Team (2008) *R: A Language and Environment for Statistical Computing*, R Foundation for Statistical Computing, Vienna, Austria, http://www.r-project.org.

Race, P. (ed.) (1999) *2000 Tips for Lecturers*, Kogan Page, London.

Race, P. (2001) *A Briefing Paper on Self, Peer and Group Assessment*, Learning and Teaching Subject Network Generic Centre Assessment Series No. 9., LTSN Generic Centre, York.

Radke-Sharpe, N. (1991) Writing as a component of statistics education. *The American Statistician*, **45**(4), 292–293.

Ramsden, P. (1992) *Learning to Teach in Higher Education*, Routledge, London.

Raymond, R. and M. Schield (2008) Numbers in the news: A survey. *ASA 2008 Proceedings of the Section on Statistical Education* [CD-ROM] 2848–2855. www.StatLit.org/pdf/2008RaymondSchieldASA.pdf (accessed 12 December 2009).

Reading, C. and Shaughnessy, J. M. (2004) Reasoning about variation, in *The Challenge of Developing Statistical Literacy, Reasoning, and Thinking*, (eds D. Ben-Zvi and J. Garfield) Kluwer, Dordrecht, The Netherlands.

Reynolds, F. (1997) Studying psychology at degree level: Would problem-based learning enhance students' experiences? *Studies in Higher Education*, **22**(3), 263–275.

Ridgway, J., Nicholson J. and McCusker, S. (2008) Reconceptualising 'statistics' and 'education'. International Association of Statistical Education (IASE), http://www.stat.auckland.ac.nz/~iase/publications/rt08/T1P5_Ridgway.pdf (accessed 12 December 2009).

Roback, P. (2003) Teaching an advanced methods course to a mixed audience. *Journal of Statistics Education*, **11**, 2, http://www.amstat.org/publications/jse/v11n2/roback.html (accessed 12 December 2009).

Roseth, C. J., Garfield, J. B. and Ben-Zvi, D. (2008) Collaboration in learning and teaching statistics. *Journal of Statistics Education*, **16**, 1, http://www.amstat.org/publications/jse/v16n1/roseth.html (accessed 12 December 2009).

Rosenthal, J. (2006) *Struck by Lightning*, Joseph Henry Press Washington.

Rossman, A. and Chance, B. (2002) A data-oriented, active learning, post-calculus introduction to statistical concepts, methods and theory, in *Proceedings of the Sixth International Conference on Teaching Statistics, Cape Town*, (ed. B. Phillips) International Statistical Institute, Voorburg, the Netherlands. http://www.stat.auckland.ac.nz/~iase/publications/1/3i2_ross.pdf.

Rumsey, D. J. (2002) Statistical literacy as a goal for introductory statistics courses. *Journal of Statistics Education*, **10**, 3, http://www.amstat.org/publications/jse/v10n3/rumsey2.html (accessed 12 December 2009).

Rumsey, D. (2005) Teaching statistics through an immersive learning environment, in *Innovations in Teaching Statistics*, (ed. J. Garfield), Mathematical Association of America, Washington, D.C., pp. 83–92.

Russell, M. B. (2005) Evaluating the weekly-assessed tutorial sheet approach to assessment: background, pedagogy and impact. *Journal for the Enhancement of Learning and Teaching*, **2**(1), 26–35.

Sadler, D. R. (1987) Specifying and promulgating achievement standards. *Oxford Review of Education*, **13**, 191–209.

Sadler, D. R. (1989) Formative assessment and the design of instructional systems. *Instructional Science*, **18**, 119–144.

Saldanha, L. A. and Thompson, P. W. (2001) Students' reasoning about sampling distributions and statistical inference, in Proceedings of the Twenty-Third Annual Meeting of the North American Chapter of the International Group for the Psychology of Mathematics Education, Snowbird, Utah, (eds R. Speiser and C. Maher), ERIC Clearinghouse, Columbus, Ohio.

Samsa, G. and Oddone, E. Z. (1994) Integrating scientific writing into a statistics curriculum: A course in statistically based scientific writing. *The American Statistician*, **48**(2), 117–119.

Sander, P., Stevenson, K., King, M. and Coates, D. (2000) University students' expectations of teaching. *Studies in Higher Education*, **25**(3), 309–323.

Savin-Baden, M. (2003) *Facilitating Problem-based Learning: Illuminating Perspectives*, Society for Research into Higher Education and Open University Press, Maidenhead.

Schield, M. and C. Schield (2007) Numbers in the news: A survey. *ASA Proceedings of the Section on Statistical Education*, [CD-ROM], pp. 2323–2328, http://www.statlit.org/pdf/2007SchieldASA.pdf (accessed 12 December 2009).

Schield, M. (2004a) *Viewpoint on Education*, http://www.augsburg.edu/ppages/~Schield/ (accessed 12 December 2009).

Schield, M. (2004b) Statistical literacy curriculum design, *IASE Curricular Development in Statistics Education Roundtable*, pp. 54–74, http://www.statlit.org/pdf/2004SchieldIASE.pdf (accessed 12 December 2009).

Schield, M. (2005) Statistical literacy and chance. 2007 *ASA Proceedings of the Section on Statistical Education*, [CD-ROM], pp. 2302–2310, http://www.statlit.org/pdf/2005SchieldASA.pdf (accessed 12 December 2009).

Schield, M. (2006) Presenting confounding and standardization graphically. *STATS Magazine*, Fall 2006, 14–18. Draft at http://www.statlit.org/pdf/2006SchieldSTATS.pdf (accessed 12 December 2009).

Schield, M. (2007) Statistical literacy: Teaching the social construction of statistics, Midwest Sociological Society, http://www.statlit.org/pdf/2007SchieldMSS.pdf (accessed 12 December 2009).

Schield, M. (2008a) Analyzing numbers in the news: A structured critical-thinking approach. National Numeracy Network Conference, http://www.statlit.org/pdf/2008SchieldNNN.pdf (accessed 12 December 2009).

Schield, M. (2008b) Quantitative literacy and school mathematics: Percentages and fractions, in *Calculation vs. Context: Quantitative Literacy and Its Implications for Teacher Education*, (eds B. L. Madison and L. A. Steen), pp. 87–107, http://www.StatLit.org/pdf/2008SchieldMAA.pdf or www.maa.org/QL/cvc/cvc-087-107.pdf (accessed 12 December 2009).

Schmeiser, C. B., and Welch, C. J. (2006) Test development, in *Educational Measurement*, 4th edn (ed. R. L. Brennan), American Council on Education/Praeger, Westport, CT, pp. 307–353.

Scriven, M. (1967) The methodology of evaluation, in *Perspectives in Evaluation*, American Educational Research Association Monograph Series on Curriculum Evaluation, No. 1, (eds R. W. Tyler *et al.*), Rand McNally, Chicago, Ill., pp. 39–83.

Seabrook, R. (2006) Is the teaching of statistical calculations helpful to students' statistical thinking?. *Psychology Learning and Teaching*, **5**, 153–161.

Shaughnessy, J. M., Watson, J., Moritz, J. and Reading, C. (1999) School mathematics students' acknowledgment of statistical variation. For the NCTM Research Precession Symposium: There's More to Life than Centers. Paper presented at the 77th Annual National Council of Teachers of Mathematics (NCTM) Conference, San Francisco, CA.

Shewhart, W. and Deming, W. (eds) (1986) [1939] *Statistical Method from the Viewpoint of Quality Control*, Dover Publications, New York.

Simonite, V., Ells, P. and Turner, W. (1998) Using IT to generate individualised coursework questions and solutions for an introductory course in statistics and probability. *CTI Maths&Stats Newsletter*, February 1998, 16–18.

Sisto, M. (2009) Can you explain that in plain English? Making statistics group projects work in a multicultural setting. *Journal of Statistics Education*, **17**, 2, http://www.amstat.org/publications/jse/v17n2/sisto.html (accessed 12 December 2009).

Snee, R (1993) What's Missing in Statistical Education? *The American Statistician* **47**, 149–154.

Snell, J. L. (1999) Using chance media to promote statistical literacy. Presented at the Joint Statistical Meeting of the American Statistical Association (ASA), http://www.statlit.org/pdf/1999SnellASA.pdf (accessed 12 December 2009).

Solomon, Y., Croft, T. and Lawson, D. (2008) Safety in numbers: Mathematics Support Centres and their derivatives as social learning spaces. Presented at the British Educational Research Association Annual Conference, Heriot Watt University, September 2008.

Starkings, S. (1997) Assessing student projects, in *The Assessment Challenge in Statistics Education*, (eds I. Gal and J. Garfield), IOS Press, Amsterdam, Netherlands, pp. 139–151. http://www.stat.auckland.ac.nz/~iase/publications/assessbk/chapter11.pdf (accessed 12 December 2009).

STARS: Creation of Statistical Resources from Real Datasets http://stars.ac.uk (accessed 12 December 2009).

StataCorp. (2003) Stata Statistical Software, release 8.2, College Station, Stata Corporation.

Statistics for the Terrified (2008) Radcliffe Medical Press Ltd., Oxford.

StatSoft, Inc. (2008) *Electronic Statistics Textbook*, http://www.statsoft.com/textbook/stathome.html (accessed 12 December 2009).

Steiner, S. and MacKay, R. J. (2009) Teaching process improvement using a virtual manufacturing environment. Submitted to *The American Statistician*.

Steen, L. A. (2001) A Mathematics and Democracy: The Case for Quantitative Literacy, Mathematical Association of America.

Stephens, R. and Nte, S. (2008) Development of an interactive visual workspace to aid the intuitive understanding of ANOVA (Analysis of Variance). Paper presented at the PLAT2008 Psychology Learning and Teaching Conference, http://www.psychology.heacademy.ac.uk/plat2008/assets/ppts/Richard_Stephensc.ppt (accessed 12 December 2009).

Stirling, W. D. (2008a) Random computer-based exercises about normal distributions, in *Proceedings of the 6th Australian Conference on Teaching Statistics, OZCOTS 2008*, (ed. B. Phillips), http://sky.scitech.qut.edu.au/~macgilli/ozcots2008/papers/OZCOTS_Stirling.pdf (accessed 12 December 2009).

Stirling, W. D. (2008b) Computer-Assisted Statistics Textbooks (CAST), release 4.0 http://cast.massey.ac.nz (accessed 12 December 2009).

Stirling, W. D. (2008c) CAST Exercises. Click ‘The e-books’ at http://cast.massey.ac.nz/collection_public.html (accessed 12 December 2009).

Strasak, A. M., Zaman, Q., Pfeiffer, K. P., Gobel, G. and Ulmer, H. (2007) Statistical errors in medical research – a review of the common pitfalls, *Swiss Medical Weekly*, **137**, 44–49.

Stromberg, A. J. and Ramanathan, S. (1996) Easy implementation of writing in introductory statistics courses. *The American Statistician*, **50**(2), 159–163.

Swets, J. A., Rubin, A. and Feurzeig, W. (1987) *Cognition, Computers, and Statistics: Software Tools for Curriculum Design*. Report No. 6447, Cambridge, MA: BBN, Inc.

Swinscow, T. D. V. (2008) *Statistics at Square One*, http://www.bmj.com/statsbk (accessed 12 December 2009).

Thelwall, M. (1999) Open access randomly generated tests: Assessment to drive learning, in *Computer Assisted Assessment in Higher Education*, (eds S. Brown, P. Race and J. Bull), Kogan Page, London.

Thelwall, M. (2000) Computer bases assessment: A versatile educational tool. *Journal of Computers and Education*, **34**, 37–49.

Topping, K. (1998) Peer assessment between students in colleges and universities. *Review of Educational Research*, **68**(3), 249–276.

Trumbo, B. E. (2002) *Learning Statistics with Real Data*, Pacific Grove, Duxbury Press, California.

Tversky, A. and Kahneman, D. (1982) Judgment under uncertainty: Heuristics and biases, in *Judgment under Uncertainty: Heuristics and Biases*. Press Syndicate of the University of Cambridge, New York, pp. 3–20, (originally published in *Science* (1974), **185**, 1124–1131).

U.S. Department of Education, Institute of Education Sciences (IES), National Center for Education Statistics (NCES). National Assessment of Educational Progress (NAEP) scores for a given subject, year, jurisdiction and grade. See http://nces.ed.gov/nationsreportcard/naepdata (accessed 12 December 2009).

Utts, J. (2003) What educated citizens should know about statistics and probability. *The American Statistician*, **57**(2), 74–79.

Utts, J. (2004) *Seeing Through Statistics*, Thomson Brooks/Cole, Belmont, CA.

Wakefield, A., Murch, S., Anthony, A. *et al.* (1998) Ileal-lymphoid-nodular hyperplasia, non-specific colitis, and pervasive developmental disorder in children. *The Lancet*, **351**(9103), 637–641.

Wallman, K. (1993) Enhancing statistical literacy: enriching our society. *Journal of the American Statistical Association*, **88**(421), 1–8.

Watson, J. M. (1997) Assessing statistical thinking using the media, in *The Assessment Challenge in Statistics Education*, (I. Gal and J. B. Garfield), IOS Press, The Netherlands, 107–121, http://www.stat.auckland.ac.nz/~iase/publications/assessbk/chapter09.pdf (accessed 12 December 2009).

Watson, J. M. (2006) *Statistical Literacy at School: Growth and Goals*, Erlbaum Associates, Mahwah, N.J.

Watson, J. and Callingham, R. (2003) Statistical literacy: A complex hierarchical construct. *Statistics Education Research Journal*, **2**(2), 3–46, http://www.stat.auckland.ac.nz/~iase/serj/SERJ2(2)_Watson_Callingham.pdf (accessed 12 December 2009).

Weldon, K. L. (2007) Assessment of a writing course in statistics, in *Proceedings of the IASE Satellite Conference on Assessing Student Learning in Statistics*, (eds B. Phillips and K. L. Weldon), International Statistical Institute, Voorburg, The Netherlands, http://www.stat.auckland.ac.nz/~iase/publications/sat07/Weldon.pdf (accessed 19 September 2009).

Wells, G., Memon, A. and Penrod, S. (2006) Eyewitness evidence: Improving its probative value. *Psychological Science in the Public Interest*, **7**(2), 45–75.

Wiggins, G. and McTighe, J. (1998) *Understanding by Design*, Association for Supervision and Curriculum Development, Alexandria, VA.

Wild, C. J. (1994) Embracing the "wider view" of statistics. *The American Statistician*, **48**(2), 163–171.

Wild, C. (2007) Virtual environments and the acceleration of experiential learning. *International Statistical Review*, **75**, 322–335.

Wild, C. J. and Pfannkuch, M. (1999) Statistical thinking in empirical enquiry (with Discussion). *International Statistical Review*, **67**(3), 223–265, http://www.stat.auckland.ac.nz/~iase/publications/isr/99.Wild.Pfannkuch.pdf (accessed 12 December 2009).

Wild, C. J., Triggs, C. and Pfannkuch, M. (1997) Assessment on a budget: Using traditional methods imaginatively, in *The Assessment Challenge in Statistics Education*, (eds I. Gal and J. B. Garfield), IOS Press, The Netherlands, pp. 205–220, http://www.stat.auckland.ac.nz/~iase/publications/assessbk/chapter16.pdf (accessed 12 December 2009).

Willis, S. A. (2002) Problem-based learning in a general psychology course. *Journal of General Education*, **51**(4), 282–291.

Wood, J. and Levy, P. (2009) Inquiry-based learning pedagogies in the arts and social sciences: Purposes, conceptions and models of practice. *Improving Student Learning: Through the Curriculum*, 2008 16th International Symposium (ed. C. Rust), the Oxford Centre for Staff and Learning Development, Oxford, pp. 128–142.

Yilmaz, M. R. (1996) The challenge of teaching statistics to non-specialists. *Journal of Statistics Education*, **4**(1), 1–9, http://www.amstat.org/publications/jse/v4n1/yilmaz.html (accessed 19 September 2009).

Yorke, M. (2003) Formative assessment in higher education: Moves towards theory and the enhancement of pedagogic practice. *Higher Education*, **45**(4), 477–501.

Young, J. (2007) Statistical errors in medical research – a chronic disease? *Swiss Medical Weekly*, **137**, 41–43.

Zeedyk, S. (2006) Detective work on Statistics Street: Teaching statistics through humorous analogy. *Psychology Learning and Teaching*, **5**(2), 97–110.

Zevenbergen, R. (2001) Peer assessment of student constructed posters: Assessment alternatives in preservice mathematics education. *Journal of Mathematics Teacher Education*, **4**, 95–113.

Index

Assessment Methods in Statistical Education Edited by Penelope Bidgood, Neville Hunt and Flavia Jolliffe